Problems and Solutions on Optics

Major American Universities Ph. D. Qualifying Questions and Solutions
Problems and Solutions on Optics

Compiled by:
The Physics Coaching Class
University of Science and
Technology of China

Refereed by:
Bai Gui-ru, Guo Guang-can

Edited by:
Lim Yung-kuo

World Scientific
Singapore • New Jersey • London • Hong Kong

Published by

World Scientific Publishing Co. Pte. Ltd.

P O Box 128, Farrer Road, Singapore 9128

USA office: 687 Hartwell Street, Teaneck, NJ 07666

UK office: 73 Lynton Mead, Totteridge, London N20 8DH

Library of Congress Cataloging-in-Publication data is available.

Major American Universities Ph.D. Qualifying Questions and Solutions

PROBLEMS AND SOLUTIONS ON OPTICS

ISBN   981-02-0438-8
         981-02-0439-6 (pbk)

Printed in Singapore by JBW Printers & Binders Pte. Ltd.

PREFACE

This series of physics problems and solutions, which consists of seven parts — Mechanics, Electromagnetism, Optics, Atomic, Nuclear and Particle Physics, Thermodynamics and Statistical Physics, Quantum Mechanics, Solid State Physics — contains a selection of 2550 problems from the graduate school entrance and qualifying examination papers of seven U. S. universities — California University Berkeley Campus, Columbia University, Chicago University, Massachusetts Institute of Technology, New York State University Buffalo Campus, Princeton University, Wisconsin University — as well as the CUSPEA and C. C. Ting's papers for selection of Chinese students for further studies in U.S.A. and their solutions which represent the effort of more than 70 Chinese physicists.

The series is remarkable for its comprehensive coverage. In each area the problems span a wide spectrum of topics while many problems overlap several areas. The problems themselves are remarkable for their versatility in applying the physical laws and principles, their uptodate realistic situations, and their scanty demand on mathematical skills. Many of the problems involve order of magnitude calculations which one often requires in an experimental situation for estimating a quantity from a simple model. In short, the exercises blend together the objectives of enhancement of one's understanding of the physical principles and practical applicability.

The solutions as presented generally just provide a guidance to solving the probelms rather than step by step manipulation and leave much to the student to work out for him/herself, of whom much is demanded of the basic knowledge in physics. Thus the series would provide an invaluable complement to the textbooks.

In editing no attempt has been made to unify the physical terms and symbols. Rather, they are left to the setters' and solvers' own preference so as to reflect the realistic situation of the usage today.

v

The present volume for Optics consists of three parts: Geometrical Optics, Physical Optics, Quantum Optics, and comprises 160 problems.

Lim Yung-kuo
Editor

INTRODUCTION

Solving problems in school work is the exercise of mental faculties, and examination problems are usually the pick of the problems in school work. Working out problems is a necessary and important aspect of the learning of Physics.

The *American University Ph. D. Qualifying Questions and Solutions* is a series of seven volumes. The subjects of the volumes and their respective referees (in parentheses) are as follows:

1. Mechanics (Qiang Yuan-qi, Gu En-pu, Cheng Jia-fu, Li Ze-hua, Yang De-tian)
2. Electromagnetism (Zhao Shu-ping, You Jun-han, Zhu Jun-jie)
3. Optics (Bai Gui-ru, Guo Guang-can)
4. Atomic, Nuclear and Particle Physics (Jin Huai-cheng, Yang Bao-zhong, Fan Yang-mei)
5. Thermodynamics and Statistical Physics (Zheng Jiu-ren)
6. Quantum Mechanics (Zhang Yong-de, Zhu Dong-pei, Fan Hong-yi)
7. Solid State Physics and Comprehensive Topics (Zhang Jia-lü, Zhou You-yuan, Zhang Shi-ling)

These books cover almost all aspects of university physics and contain 2550 problems, most of which are solved in detail.

These problems have been carefully chosen from a collection of 3100 problems some of which came from the China-U.S.A. Physics Examination and Application Programme and Ph.D. Qualifying Examination on Experimental High Energy Physics sponsored by Chao Chong Ting, while the others from the graduate preliminary or qualifying examination questions of the following seven top American universities during the last decade: Columbia University, University of California at Berkeley, Massachusetts Institute of Technology, University of Wisconsin, University of Chicago, Princeton University, State University of New York at Buffalo.

In general, examination problems on physics in American universities do not involve too much mathematics. Rather, they can be categorized into the following three types. Many of the problems that involve the various frontier subjects and overlapping domains of science have been selected by the professors directly from their own research and show a "modern style". Some of the problems involve a wide field and require a quick mind to analyse, while the others are often simple to solve but are practical and require a full "touch of physics." We think it reasonable to take these problems as a reflection, to some extent, of the characteristics of American science and culture, as well as the tenet of American education.

This being so, we believe it worthwhile to collect and solve these problems and then introduce them to the students and teachers, even though the effort involved is formidable. Nearly a hundred teachers and graduate students took part in this time-consuming task.

There are 160 problems in this volume, which is divided into three parts: part I consists of 41 problems in geometric optics, part II consists of 89 problems in wave optics, part III consists of 30 problems in quantum optics.

The depth of knowledge involved in solving these problems is not beyond the contents of common textbooks on optics used in colleges and universities in China, although the scope of the knowledge and techniques needed in solving some of the problems go beyond what we are usually familiar with. Furthermore, some new scientific research results (e.g. some newly developed lasers) are introduced in the problems. This will not only enhance the understanding of the established theories and knowledge, but also encourage the interaction between teaching and research which cannot but enliven academic thoughts and excite the mind.

The physicists who contributed to solving the problems in this volume are Shi De-xiu, Yao Kun, Lu Hong-jun, Chen Xiang-li, Gu Chun, Han Wen-hai and Wu Zhi-qiang. The initial translation from Chinese into English was carried out by Xuan Zhi-hua. Some revisions have been made in this English edition by the compilers, the translator and the editor.

CONTENTS

Preface   v

Introduction   vii

Part 1   Geometrical Optics (1001–1041)   1

Part 2   Wave Optics (2001–2089)   51

Part 3   Quantum Optics (3001–3030)   151

PART 1 GEOMETRICAL OPTICS

PART I GEOMETRICAL OPTICS

1001

A rainbow is produced by:

(a) refraction of sunlight by water droplets in the atmosphere.
(b) reflection of sunlight by clouds.
(c) refraction of sunlight in the human eye.

(CCT)

Solution:

The answer is (a).

1002

A horizontal ray of light passes through a prism of index 1.50 and apex angle 4° and then strikes a vertical mirror, as shown in the figure. Through what angle must the mirror be rotated if after reflection the ray is to be horizontal?

(Wisconsin)

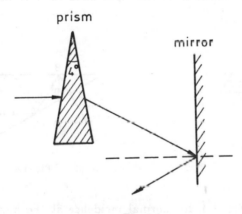

Fig. 1.1

Solution:

As the apex angle is very small ($\alpha = 4°$), the angle of deviation δ can be obtained approximately:

$$\delta = (n - 1)\alpha = (1.5 - 1) \times 4° = 2° .$$

From Fig. 1.2 we see that if the reflected ray is to be horizontal, the mirror must be rotated clockwise through an angle γ given by

$$\gamma = \frac{\delta}{2} = 1° .$$

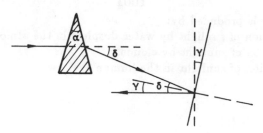

Fig. 1.2

1003

A narrow beam of light is incident on a $30° - 60° - 90°$ prism as shown (Fig. 1.3). The index of refraction of the prism is $n = 2.1$. Show that the entire beam emerges either from the right-hand face, or back along the incident path.

(Wisconsin)

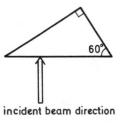

incident beam direction

Fig. 1.3

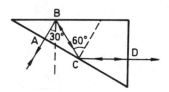

Fig. 1.4

Solution:

As seen from Fig. 1.4, for normal incidence at the bottom face the angle of incidence at B is 30°, and that at C is 60°, both of which are larger than the critical angle of the prism,

$$\theta_c = \sin^{-1}\frac{1}{n} = 28°26' \ .$$

Hence the ray is totally reflected at B and C. Also, the ray is partially reflected back at the bottom and the right-hand faces. Therefore, the entire beam emerges either from the right-hand face, or back along the incident path.

1004

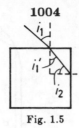

Fig. 1.5

A glass cube has a refractive index of 1.5. A light beam enters the top face obliquely and then strikes the side of the cube. Does light emerge from this side? Explain your answer.

(*Wisconsin*)

Solution:

Assume that the angles of incidence and refraction at the top face are i_1 and i_1' respectively. According to Snell's law of refraction,

$$\sin i_1 = n \sin i_1' .$$

From the geometry (see Fig. 1.5) $i_1' + i_2 = 90°$, where i_2 is the angle of incidence at the right side. Thus, we have

$$\sin i_1 = n \cos i_2 ,$$

or

$$i_2 = \cos^{-1}\left(\frac{\sin i_1}{n}\right) .$$

When $i_1 = 90°$, i_2 has the minimum value

$$i_2 = \cos^{-1}\frac{1}{1.5} = 48°10' > i_c = 42° \quad (\text{critical angle}) .$$

Hence no light emerges from this side.

1005

A glass rod of rectangular cross-section is bent into the shape shown in the Fig. 1.6. A parallel beam of light falls perpendicularly on the flat surface A. Determine the minimum value of the ratio R/d for which all light entering the glass through surface A will emerge from the glass through surface B. The index of refraction of the glass is 1.5.

(*Wisconsin*)

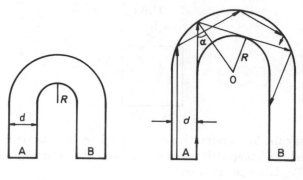

Fig. 1.6   Fig. 1.7

Solution:

Consider the representative rays shown in Fig. 1.7. A ray entering the glass through surface A and passing along the inner side of the rod will be reflected by the outer side with the smallest angle α, at which the reflected ray is tangent to the inner side. We have to consider the conditions under which the ray will undergo total internal reflection before reaching B.

If $\alpha > \theta_c$, the critical angle, at which total internal reflection occurs, all the incident beam will emerge through the surface B. Hence we require

$$\sin \alpha > \frac{1}{n} \ .$$

The geometry gives

$$\sin \alpha = \frac{R}{(R + d)} \ .$$

Therefore

$$\frac{R}{R + d} \geq \frac{1}{n} \ ,$$

or

$$\left(\frac{R}{d}\right)_{\min} = \frac{1}{n - 1} = \frac{1}{1.5 - 1} = 2 \ .$$

1006

A small fish, four feet below the surface of Lake Mendota is viewed through a simple thin converging lens with focal length 30 feet. If the lens is 2 feet above the water surface (Fig. 1.8), where is the image of the fish seen by the observer? Assume the fish lies on the optical axis of the lens and that $n_{air} = 1.0, n_{water} = 1.33$.

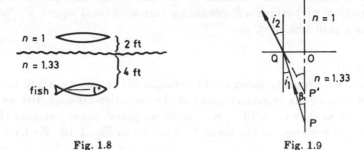

Fig. 1.8   Fig. 1.9

Solution:

An object at P in water appears to be at P′ as seen by an observer in air, as Fig. 1.9 shows. The paraxial light emitted by P is refracted at the water surface, for which

$$1.33 \sin i_1 = \sin i_2 .$$

As i_1, i_2 are very small, we have the approximation $1.33 i_1 = i_2$. Also,

$$i_2 = \alpha \approx \frac{\overline{OQ}}{\overline{OP'}}, \quad i_1 = \beta \approx \frac{\overline{OQ}}{\overline{QP}} .$$

Hence, we have

$$\overline{OP'} = \frac{1}{1.33} \cdot \overline{OP} = 3 \text{ ft} .$$

Let the distance between the apparent location of the fish and the center of the lens be u, then

$$u = 2 + \overline{OP'} = 5 \text{ ft} .$$

From $\frac{1}{f} = \frac{1}{u} + \frac{1}{v}$, we have

$$v = -6 \text{ ft} .$$

Therefore, the image of the fish is still where the fish is, four feet below the water surface.

1007

The index of refraction of glass can be increased by diffusing in impurities. It is then possible to make a lens of constant thickness. Given a disk of radius a and thickness d, find the radial variation of the index of refraction $n(r)$ which will produce a lens with focal length F. You may assume a thin lens $(d \ll a)$.

(Chicago)

Solution:

Let the refractive index of the material of the disk be n and the radial distribution of the refractive index of the impurity-diffused disk be represented by $n(r)$, with $n(0) = n_0$. Incident plane waves entering the lens refract and converge at the focus F as shown in Fig. 1.10. We have

$$[n(r) - n_0]d = -\sqrt{F^2 + r^2} + F ,$$

i.e.,

$$n(r) = n_0 - \frac{\sqrt{F^2 + r^2} - F}{d} .$$

For $F \gg r$, we obtain

$$n(r) = n_0 - \frac{r^2}{2dF} .$$

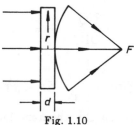

Fig. 1.10

1008

The index of refraction of air at 300 K and 1 atmosphere pressure is 1.0003 in the middle of the visible spectrum. Assuming an isothermal atmosphere at 300 K, calculate by what factor the earth's atmosphere would have to be more dense to cause light to bend around the earth with the earth's curvature at sea level. (In cloudless skies we could then watch sunset all night, in principle, but with an image of the sun drastically compressed vertically.) You may assume that the index of refraction n has the property that $n - 1$ is proportional to the density. (Hint: Think of Fermat's Principle.) The $1/e$ height of this isothermal atmosphere is 8700 metres.

(*UC, Berkeley*)

Solution:

We are given that

$$n(r) - 1 = \rho e^{-\frac{r-R}{8700}} \ ,$$

where $R = 6400 \times 10^3$ m is the earth's radius and ρ is the density coefficient of air. Then

$$n(r) = 1 + \rho e^{-\frac{r-R}{8700}} \ , \tag{1}$$

$$\frac{dn(r)}{dr} = n'(r) = -\frac{1}{8700} \rho e^{-\frac{r-R}{8700}} \ . \tag{2}$$

It is also given that air is so dense that it makes light bend around the earth with the earth's curvature at sea level, as shown in Fig. 1.11.

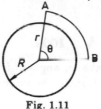

Fig. 1.11

The optical path length from A to B is

$$l = n(r) r \theta \ .$$

According to Fermat's Principle, the optical path length from A to B should be an extremum. Therefore,

$$\frac{dl}{dr} = [n'(r)r + n(r)]\theta = 0 \ ,$$

i.e.,

$$n'(r) = \frac{-n(r)}{r} \ . \tag{3}$$

Substituting (3) into (2) yields

$$\frac{1}{8700} \rho e^{-\frac{r-R}{8700}} = \frac{n(r)}{r} \ . \tag{4}$$

At sea level, $r = R = 6400 \times 10^3$ m. This with (1) and (4) yields

$$\frac{\rho \times 6400 \times 10^3}{8700} = 1 + \rho \,,$$

giving

$$\rho = 0.00136 \,.$$

At sea level, i.e., at 300 K and 1 atmosphere pressure, $n_0 - 1 = \rho_0 = 0.0003$. Therefore

$$\frac{\rho}{\rho_0} = 4.53 \,.$$

Thus only if the air were 4.53 times as dense as the real air would light bend around the earth with the earth's curvature at sea level.

1009

Incident parallel rays make an angle of 5° with the axis of a diverging lens −20 cm in focal length. Locate the image.

(*Wisconsin*)

Solution:

$20 \times \tan 5° = 1.75$ cm. The image is a virtual point image in the focal plane 1.75 cm off the optical axis.

1010

A thin lens with index of refraction n and radii of curvature R_1 and R_2 is located between 2 media with indices of refraction n_1 and n_2 as shown (Fig. 1.12). If S_1 and S_2 are the object and image distances respectively, and f_1 and f_2 the respective focal lengths, show that

$$\frac{f_1}{S_1} + \frac{f_2}{S_2} = 1 \,.$$

(*Wisconsin*)

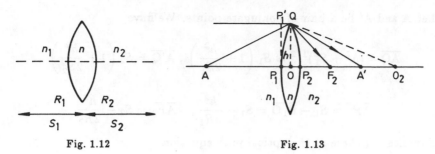

Fig. 1.12   Fig. 1.13

Solution:

First we study the relation between f_1, f_2 and R_1, R_2, n_1, n_2, n. As Fig. 1.13 shows, a ray parallel to the axis is refracted at Q and crosses the axis at the second focal point F_2; a ray along the axis passes through F_2 also. As the optical lengths of the two rays are equal, we have

$$n_1\overline{P_1'Q} + n_2\overline{QF_2} = n\overline{P_1P_2} + n_2\overline{P_2F_2} \ . \tag{1}$$

As $\overline{QO_2} = \overline{P_1O_2} = R_1$, we have

$$\overline{P_1O} - R_1 - \overline{OO_2} - R_1 - \sqrt{R_1^2 - h^2} \approx \frac{h^2}{2R_1} \ , \ \left(\frac{h}{R_1} \ll 1\right) \ .$$

In the same way, we have

$$\overline{P_2O} \approx \frac{h^2}{2R_2} \ .$$

Hence $\overline{P_1P_2} = \frac{h^2}{2}\left(\frac{1}{R_1} + \frac{1}{R_2}\right)$. As $\overline{P_1'Q} = \overline{P_1O}$,

$$\overline{QF_2} = (h^2 + f_2^2)^{\frac{1}{2}} \approx f_2\left(1 + \frac{h^2}{2f_2^2}\right) \ ,$$

$$\overline{P_2F_2} = f_2 - \overline{P_2O} = f_2 - \frac{h^2}{2R_2} \ ,$$

substituting into (1) yields

$$\frac{n_2}{f_2} = \frac{n - n_1}{R_1} + \frac{n - n_2}{R_2} \ . \tag{2}$$

Similarly we obtain

$$\frac{n_1}{f_1} = \frac{n - n_1}{R_1} + \frac{n - n_2}{R_2} \ . \tag{3}$$

Let A and A' be a pair of conjugate points. We have

$$\overline{AQ} = (h^2 + S_1^2)^{1/2} \approx S_1 \left(1 + \frac{h^2}{2S_1^2}\right), \ \overline{A'Q} = S_2 \left(1 + \frac{h^2}{2S_2^2}\right),$$

$$\overline{AP_1} = S_1 - \overline{P_1O} = S_1 - \frac{h^2}{2R_1}, \quad \overline{AP_2} = S_2 - \frac{h^2}{2R_2}.$$

Substituting these in the optical path equation

$$n_1\overline{AQ} + n_2\overline{A'Q} = n_1\overline{AP_1} + n\overline{P_1P_2} + n_2\overline{A'P_2}$$

yields

$$\frac{n_1}{S_1} + \frac{n_2}{S_2} = \frac{n - n_1}{R_1} + \frac{n - n_2}{R_2}. \tag{4}$$

Combining (2), (3) and (4) yields

$$\frac{f_1}{S_1} + \frac{f_2}{S_2} = 1.$$

1011

A line object 5 mm long is located 50 cm in front of a camera lens. The image is focussed on the film plate and is 1 mm long. If the film plate is moved back 1 cm the width of the image blurs to 1 mm wide. What is the F-number of the lens?

(*Wisconsin*)

Solution:

Substituting $u = 50$ cm and $\frac{v}{u} = \frac{1}{5}$ in the Gaussian lens formula

$$\frac{1}{u} + \frac{1}{v} = \frac{1}{f}$$

gives $f = 8.33$ cm, $v = 10$ cm. From the similar triangles in Fig. 1.14 we have

$$\frac{D}{v} = \frac{0.1}{1},$$

or $D = 0.1v = 1$ cm. Therefore, $F = f/D = 8.33$.

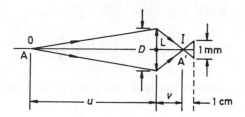

Fig. 1.14

1012

For a camera lens, "depth of field" is how far a point object can be away from the position where it would be precisely in focus and still have the light from it fall on the film within a "circle of confusion" of some diameter, say l. For a given picture derive a relation for the depth of field, Δq, as a function of the object distance q, the focal length of the lens, the f stop and l. (You may consider the object distance to be much larger than the focal length.)

(*Wisconsin*)

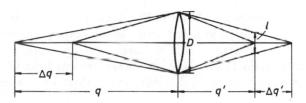

Fig. 1.15

Solution:

The Gaussian lens formula

$$\frac{1}{q} + \frac{1}{q'} = \frac{1}{f}$$

gives $\frac{dq'}{dq} = -\left(\frac{q'}{q}\right)^2$. Thus for a small deviation of the object distance (depth of field), Δq, the deviation of the image distance, is

$$|\Delta q'| \approx |\Delta q| \left(\frac{q'}{q}\right)^2 .$$

By geometry (Fig. 1.15),

$$\frac{\Delta q'}{l} = \frac{(q' + \Delta q')}{D} \approx \frac{q'}{D},$$

where D is the diameter of the lens. As $q \gg f, q' \approx f$, we obtain

$$\Delta q \approx \frac{lq'}{D}\left(\frac{q}{q'}\right)^2 \approx \frac{lf}{D}\left(\frac{q}{f}\right)^2 = \frac{l}{F}\left(\frac{q}{f}\right)^2,$$

where $F = \frac{D}{f}$ is the f stop of the lens.

1013

Illustrate by a sketch the position and orientation of the image of the 3-arrow object (Fig. 1.16). The length of each arrow is 1/2 unit and the point O is located 3/2 F from the center of the convex lens, (F=1 unit). Work out the length of the image arrows.

(Wisconsin)

Solution:

The image of arrow a is shown in Fig. 1.17. From the geometry we get the length of the image, which is 1 unit. By symmetry, the length of the image arrow b is the same as that of itself. The arrowhead of arrow c is just at the focal point F; therefore, its image extends from O′ to infinity. Fig. 1.16 shows the positions and orientation of these images arrows.

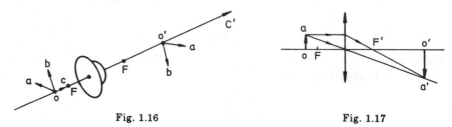

Fig. 1.16   Fig. 1.17

1014

A 55 year old man can focus objects clearly from 100 cm to 300 cm. Representing the eye as a simple lens 2 cm from the retina,

(a) What is the focal length of the lens at the far point (focussed at 300 cm)?

(b) What is the focal length of the lens at the near point (focussed at 100 cm)?

(c) What strength lens (focal length) must he wear in the lower part of his bifocal eyeglasses to focus at 25 cm?

(*Wisconsin*)

Solution:

(a) $\frac{1}{f_{far}} = \frac{1}{u_{far}} + \frac{1}{v}$, $u_{far} = 300$ cm, $v = 2$ cm.
Solving the equation yields $f_{far} = 1.987$ cm.

(b) $\frac{1}{f_{near}} = \frac{1}{u_{near}} + \frac{1}{v}$, $u_{near} = 100$ cm, $v=2$ cm.
Solving the equation yields $f_{near} = 1.961$ cm.

(c) In order to see an object at 25 cm clearly, the dioptric power should be

$$\Phi = \frac{1}{u} + \frac{1}{v} = \frac{1}{0.25} + \frac{1}{0.02} = 54 \quad \text{diopters},$$

and the combined power is the sum of the powers of the eye and the glasses,

$$\Phi = \Phi_{eye} + \Phi_{glasses}.$$

Thus $\Phi_{glasses} = \Phi - \Phi_{eye} - 54 - \frac{1}{0.01961} = 3$ diopters $\left(\Phi_{eye} = \frac{1}{f_{near}}\right)$. He must wear $300°$ far-sighted eyeglasses. The corresponding focal length is

$$f_{glasses} = \frac{1}{\Phi_{glasses}} = \frac{1}{3} \text{ m} = 33.3 \text{ cm}.$$

1015

A retro-reflector is an optical device which reflects light back directly whence it came. The most familiar retro-reflector is the reflecting corner cube, but recently the 3M Company invented "Scotchlite" spheres.

(a) Calculate the index of refraction n and any other relevant parameters which enable a sphere to retro-reflect light.

(b) Sketch how you think Scotchlite works, and discuss qualitatively the factors which might determine the reflective efficiency of Scotchlite.

(*UC, Berkeley*)

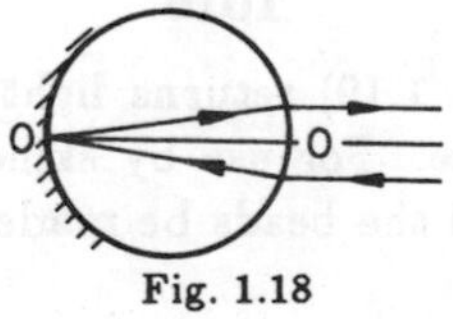

Fig. 1.18

Solution:

(a) The "Scotchlite" sphere is a ball of index of refraction n, whose rear semi-spherical interface is a reflecting surface. The focal length in the image space, f, for a single refractive interface is given by

$$f = \frac{nr}{n-1} \, ,$$

where r is the radius of the sphere. The index of refraction of air is unity. The index of refraction of the glass is chosen so that the back focal point of the front semi-spherical interface coincides with the apex of the rear semi-spherical interface (see Fig. 1.18), i.e.,

$$f = 2r \, .$$

Hence $n = 2$.

(b) The rear semi-spherical interface of the "Scotchlite" sphere reflects the in-coming light partially, its retro-reflectance efficiency η being given by

$$\eta = T^2 R \, ,$$

where T is the transparency of the front semi-spherical interface at which light is refracted twice, being

$$T = \frac{4n}{(n+1)^2} = 0.89 \, ,$$

and R is the reflectance of the rear semi-spherical interface. Here we have assumed that no absorption occurs. For silver coating, $R = 0.95$, we have

$$\eta = 0.89^2 \times 0.95 = 75\% \, .$$

1016

A beaded screen (Fig. 1.19) returns light back to the source if light focusses on its back surface. For use by skindivers in water $\left(n = \frac{4}{3}\right)$, of what index material should the beads be made ideally?

(Wisconsin)

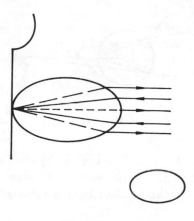

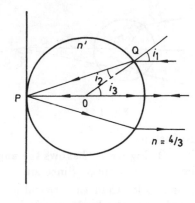

Fig. 1.19   Fig. 1.20

Solution:

Let the required index of refraction be n', (see Fig. 1.20). If a pencil of paraxial rays which are parallel to the axis, OP, strikes at P, the reflected rays will return back parallel to OP. Snell's law of refraction $n' \sin i_2 = n \sin i_1$ for small angles i_1 and i_2 becomes $n' i_2 = n i_1$. As $i_3 = i_1$, $i_3 = 2i_2$, we have $i_1 = 2i_2$. Therefore, $n' = 2n = \frac{8}{3}$.

1017

A ray of light enters a spherical drop of water of index n as shown (Fig. 1.21).

(1) What is the angle of incidence α of the ray on the back surface? Will this ray be totally or partially reflected?

(2) Find an expression for the angle of deflection δ.

(3) Find the angle ϕ which produces minimum deflection.

$$\left(\text{Hint}: \quad \frac{d \sin^{-1} x}{dx} = \frac{1}{\sqrt{1-x^2}} \right) .$$

(CUSPEA)

Fig. 1.21   Fig. 1.22

Solution:

(1) Figure 1.22 shows the angles ϕ, α, $\phi - \alpha$, x, δ. α is given by Snell's law $n \sin \alpha = \sin \phi$. Since $\sin \alpha = \frac{1}{n} \sin \phi < \frac{1}{n}$, i.e., the angle of incidence α is smaller than the critical angle $\sin^{-1} \frac{1}{n}$, the incident light is reflected partially at the back spherical surface.

(2) As $\alpha = (\phi - \alpha) + x$, or $x = 2\alpha - \phi$, we have $\delta = \pi - 2x = \pi - 4\alpha + 2\phi$.

(3) For minimum deflection, we require $\frac{d\delta}{d\phi} = -4\frac{d\alpha}{d\phi} + 2 = 0$, or $\frac{d\alpha}{d\phi} = \frac{1}{2}$. As

$$\alpha = \sin^{-1}\left(\frac{1}{n}\sin\phi\right) ,$$

we have $\frac{d\alpha}{d\phi} = \frac{1}{n}\frac{\cos\phi}{\cos\alpha}$ and the above gives

$$1 - \frac{1}{n^2}\sin^2\phi = \frac{4}{n^2}\cos^2\phi ,$$

or

$$1 = \frac{1}{n^2} + \frac{3}{n^2}\cos^2\phi ,$$

giving

$$\cos^2\phi = \frac{n^2 - 1}{3} .$$

1018

It was once suggested that the mirror for an astronomical telescope could be produced by rotating a flat disk of mercury at a prescribed angular velocity ω about a vertical axis.

(a) What is the equation of the reflection (free) surface so obtained?

(b) How fast must the disk be rotated to produce a 10 cm focal length mirror?

(*Wisconsin*)

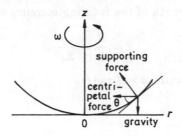

Fig. 1.23

Solution:

(a) Owing to the symmetry of the reflecting surface, we need only to consider the situation in a meridian plane. Let the equation of the reflecting surface be represented by (Fig. 1.23)

$$z = f(r) \ .$$

From dynamical considerations we have for a volume element dV

$$S \cos \theta = \rho g dV$$

$$S \sin \theta = \rho \omega^2 r dV$$

where ρ and S represent the density of mercury and the supporting force on dV respectively. Thus

$$\tan \theta = \frac{\omega^2 r}{g} \ .$$

As

$$\tan \theta = \frac{dz}{dr} ,$$

we obtain the differential equation for the reflecting surface,

$$dz = \frac{\omega^2 r dr}{g} \ .$$

Integrating and putting $z = 0$ for $r = 0$, we have

$$z = \frac{\omega^2 r^2}{2g} \ .$$

For a spherical surface of radius R,

$$z = R - \sqrt{R^2 - r^2} \approx \frac{r^2}{2R}$$

if $r \ll R$. The focal length of the rotating mercury surface is then

$$f = \frac{R}{2} = \frac{g}{2\omega^2} .$$

For $f = 10$ cm, we require $\omega = 7$ rad/s.

1019

A spherical concave shaving mirror has a radius of curvature of 12 inches. What is the magnification when the face is 4 inches from the vertex of the mirror? Include a ray diagram of the image formation.

(*Wisconsin*)

Solution:

The focal length of the spherical concave mirror is $f = \frac{r}{2} = 6$ inches, the object distance is $u = 4$ inches.

Using the formula $\frac{1}{f} = \frac{1}{u} + \frac{1}{v}$, we get the image distance $v = -12$ inches.

The minus sign signifies that the image is virtual. The magnification is

$$m = \left| \frac{v}{u} \right| = 3 .$$

The ray diagram of the image formation is shown in Fig. 1.24.

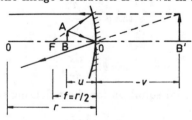

Fig. 1.24

1020

(a) A curved mirror brings collimated light to focus at $x = 20$ cm.

(b) Then it is filled with water $n = \frac{4}{3}$ and illuminated through a pinhole in a white card (Fig. 1.25). A sharp image will be formed on the card at what distance, X?

(*Wisconsin*)

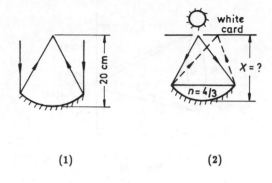

(1)   (2)

Fig. 1.25

Solution:

From (a), we find that the focal length of the mirror is $f_a = 20$ cm (in air).

Suppose the focal length is f_b when the mirror is filled with water. When paraxial rays are refracted at a plane surface, the object distance y and the image distance y' are related by

$$y' = \frac{n'y}{n},$$

where n and n' are the refractive indices of the two media (see **1006**). As now $y = f_a, n' = 1, n = 1.33$, we have $f_b = y' = \frac{f_a}{n} = \frac{20}{\left(\frac{4}{3}\right)} = 15$ cm. For a concave mirror, if the object distance is equal to the image distance then it is twice the focal length, i.e.,

$$X = 2f_b = 30 \quad \text{cm}.$$

1021

Given two identical watch glasses glued together, the rear one silvered. Using autocollimation as sketched (Fig. 1.26), sharp focus is obtained for $L = 20$ cm. Find L for sharp focus when the space between the glasses is subsequently filled with water, $n = \frac{4}{3}$.

(Wisconsin)

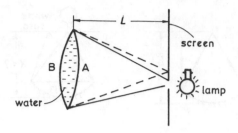

Fig. 1.26

Solution:

With air between the glasses, only the silvered watch glass reflects and converges the rays to form an image, i.e., the system acts as a concave mirror. The formula for a concave mirror

$$\frac{1}{u} + \frac{1}{v} = \frac{2}{r}$$

gives for $u = v = 20$ cm, $r = 20$ cm.

With water between the glasses, the incident light is refracted twice at A and reflected once at B before forming the final image. Note that the first image formed by A falls behind the mirror B and becomes a virtual object to B. Similarly the image formed by B is a virtual object to A. We therefore have

$$\frac{1}{L} + \frac{n}{x} = \frac{n-1}{20},$$

$$-\frac{1}{x} + \frac{1}{y} = \frac{1}{10},$$

$$-\frac{n}{y} + \frac{1}{L} = \frac{n-1}{20},$$

which yield $L = 12$ cm.

Thus, a sharp image will be formed at $L = 12$ cm.

1022

An object is placed 10 cm in front of a convering lens of focal length 10 cm. A diverging lens of focal length -15 cm is placed 5 cm behind the converging lens (Fig. 1.27). Find the position, size and character of the final image.

(*Wisconsin*)

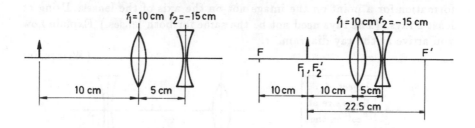

Fig. 1.27   Fig. 1.28

Solution:

For the first lens, the object distance is $u_1 = 10$ cm. As $f_1 = 10$ cm also, the image distance is $v_1 = \infty$. Then for the second lens, the object distance is $u_2 = \infty$. As $f_2 = -15$ cm, $v_2 = -15$ cm. Hence, the image coincides with the object.

Now consider the size and character of the final image. The focal length of the combined lens system is $f = -\frac{f_1 f_2}{\Delta}$, with $\Delta = d - f_1 - f_2$. As $d = 5$ cm, $f_1 = 10$ cm, $f_2 = -15$ cm, we have $f = 15$ cm; also, as shown in Fig. 1.28,

$$\overline{F_1 F} = \frac{f_1^2}{\Delta} = 10 \text{ cm},$$

$$\overline{F_2' F'} = \frac{-f_2^2}{\Delta} = -22.5 \text{ cm}.$$

Using Newton's formula, $xx' = f^2$, we have for $x = -10$ cm

$$x' = -22.5 \text{ cm}.$$

The negative sign indicates that the image is to the left of F'. The magnification is $m = \frac{-f}{x} = 1.5$. Then the image is upright, virtual and magnified 1.5 times.

1023

As shown in Fig. 1.29, the image which would be cast by the converging lens alone has a distance of 0.5 cm between top and bottom. Calculate the position and size of the final image. Draw a ray diagram showing image formation for a point on the image not on the axis of the lenses. Using at least 2 rays. (The rays need not be the same for both lenses.) Explain how you arrive at the ray diagram.

(*Wisconsin*)

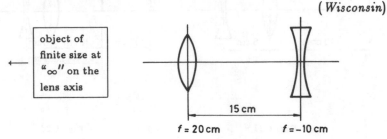

Fig. 1.29

Solution:

An object infinite distance away finds its image on the back focal plane of the converging lens, i.e., at $20 - 15 = 5$ cm behind the diverging lens. The Gaussian lens formula

$$\frac{1}{u} + \frac{1}{v} = \frac{1}{f}$$

applied to the diverging lens then gives for $u = -5$ cm and $f = -10$ cm, $v = 10$ cm. The lateral magnification is

$$m = \frac{h}{h'} = \frac{10}{5} = 2 \,.$$

Hence the size of the final image is

$$0.5 \times 2 = 1 \text{ cm} \,.$$

Fig. 1.30 gives the ray diagram:

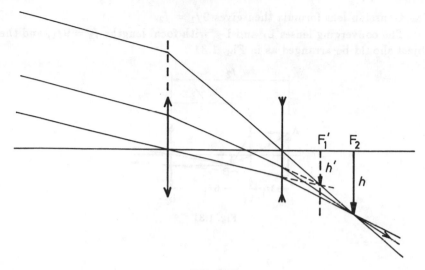

Fig. 1.30

1024

Find and describe a combination of two converging lenses which produces an inverted, virtual image at the position of the object and of the same size. By a sketch of image formation make it clear that your scheme will work.

(*Wisconsin*)

Solution:

L_1, L_2 represent two converging lenses, and $f_1, u_1, v_1, f_2, u_2, v_2$ represent the focal lengths, object distances and image distances in the sequential formation of images, respectively, as shown in Fig. 1.31.

Consider L_2 first. The object of L_2 must be within the focal length to form a virtual image. If B, the image of L_1, or the object of L_2, is at a distance of $\frac{f_2}{2}$ in front of L_2, the final image C would be at a distance of $v_2 = -f_2$ in front of L_2, where the real object A is placed, and would be twice as large as B.

Now, we have to insert L_1 between A and B so that C, which is twice the size of B, is as large as A. Thus $u_1 = 2v_1$. Furthermore, as $f_2 = u_1 + v_1 + u_2$ we obtain

$$u_1 = \frac{f_2}{3}, \ v_1 = \frac{f_2}{6}.$$

The Gaussian lens formula then gives $9f_1 = f_2$.

The converging lenses L_1 and L_2, with focal lengths $f_2 = 9f_1$, and the object should be arranged as in Fig. 1.31.

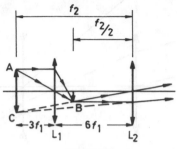

Fig. 1.31

1025

Two positive thin lenses L_1 and L_2 of equal focal length are separated by a distance of half their focal length (Fig. 1.32):

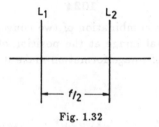

Fig. 1.32

(a) Locate the image position for an object placed at distance $4f$ to the left of L_1.

(b) Locate the focal points of this lens combination treated as a single thick lens.

(c) Locate the principal planes for this lens combination treated as a single thick lens.

(*Wisconsin*)

Solution:

(a) Let u_1, v_1, u_2, v_2 represent object and image distances for L_1 and L_2 respectively. As $u_1 = 4f, \frac{1}{v_1} = \frac{1}{f_1} - \frac{1}{u_1} = \frac{3}{4f}, v_1 = \frac{4}{3}f$. Then as $u_2 = \frac{f}{2} - v_1 = -\frac{5}{6}f$, we have $\frac{1}{v_2} = \frac{1}{f} - \frac{1}{u_2} = \frac{11}{5f}$ or $v_2 = \frac{5f}{11}$.

Hence the final image is at $\frac{5f}{11}$ right of L_2.

(b) Consider a beam of parallel light falling on L, from the left, then $v_1 = f$ and

$$u_2 = \frac{f}{2} - v_1 = -\frac{f}{2}.$$

Therefore

$$\frac{1}{v_2} = \frac{1}{f} + \frac{2}{f} = \frac{3}{f}, \text{ or } v_2 = \frac{1}{3}f.$$

By symmetry, the two local points of the lens combination lie $\frac{f}{3}$ to the left of L_1 and $\frac{f}{3}$ to the right of L_2.

(c) On the principal planes, the lateral magnification = 1. By symmetry, the image of an object on the left principal plane formed by L_1 must coincide with the image of the same object on the right principal plane formed by L_2, and both must be at the mid-point between the two lenses.

Let the left principal plane be at x left of L_1. Then $u_1 = x, v_1 = \frac{1}{2}(\frac{f}{2}) = \frac{1}{4}f$, and from $\frac{1}{f} = \frac{1}{u_1} + \frac{1}{v_1}$ we obtain $x = -\frac{1}{3}f$.

Therefore, the two principal planes are at $\frac{f}{3}$ to the right of L_1 and $\frac{f}{3}$ to the left of L_2.

1026

A self-luminous object of height h is 40 cm to the left of a converging lens with a focal length of 10 cm. A second converging lens with a focal length of 20 cm is 30 cm to the right of the first lens.

(a) Calculate the position of the final image.

(b) Calculate the ratio of the height of the final image to the height h of the object.

(c) Draw a ray diagram, being careful to show just those rays needed to locate the final image starting from the self-luminous object.

(*Wisconsin*)

Solution:

(a) From $\frac{1}{f_1} = \frac{1}{u_1} + \frac{1}{v_1}$, where $f_1 = 10$ cm, $u_1 = 40$ cm, we obtain $v_1 = 13\frac{1}{3}$ cm.

From $\frac{1}{f_2} = \frac{1}{u_2} + \frac{1}{v_2}$, where $f_2 = 20$ cm, $u_2 = (30 - \frac{40}{3})$ cm, we obtain $v_2 = -100$ cm, i.e., the final image is 100 cm to the left of the second lens.

(b) The ratio of the final image height to the height h of the object can be calculated from

$$m = m_2 \times m_1 = \left(\frac{v_2}{u_2}\right)\left(\frac{v_1}{u_1}\right) = -2.$$

The minus sign signifies an inverted image.

(c) Fig. 1.33 shows the ray diagram.

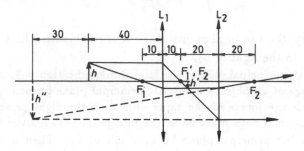

Fig. 1.33

1027

(a) What is the minimum index of refraction for the plastic rod (Fig. 1.34) which will insure that any ray entering at the end will always be totally reflected in the rod?

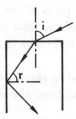

Fig. 1.34

(b) Draw a ray diagram showing image formation for the lenses and object as shown (Fig. 1.35). Choose the focal length of the second lens so that the final image will be at infinity. Use the arrow head as the object and draw at least 2 rays to show image formation. Explain briefly how you arrive at the ray diagram.

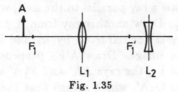

Fig. 1.35

Do all the refracting at planes 1 and 2.

(c) How would you change the position of the second (diverging) lens to make the combination a telephoto lens?

(*Wisconsin*)

Solution:

(a) As shown in Fig. 1.36, a ray would be totally reflected in the rod if

$$\theta_3 > \alpha = \sin^{-1}\left(\frac{1}{n}\right),$$

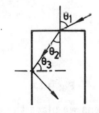

Fig 1.36

where α is the critical angle, n is the index of refraction of the plastic, assuming the index of refraction of air to be unity. Thus we require that

$$\theta_2 = \frac{\pi}{2} - \theta_3 < \frac{\pi}{2} - \sin^{-1}\left(\frac{1}{n}\right),$$

or

$$\sin\theta_1 = n\sin\theta_2 < n\sin\left(\frac{\pi}{2} - \sin^{-1}\left(\frac{1}{n}\right)\right)$$

$$= n\cos\left(\sin^{-1}\frac{1}{n}\right).$$

As $\theta_1 \leq \frac{\pi}{2}$, or $\sin\theta_1 \leq 1$, we require $n\cos(\sin^{-1}(\frac{1}{n})) > 1$ for all rays to be totally reflected in the rod. Hence the condition for total reflection is

$$\sin^{-1}\left(\frac{1}{n}\right) < \cos^{-1}\left(\frac{1}{n}\right), \text{ or } n > \sqrt{2} = 1.414.$$

(b) First let us consider the effect of the first lens, L_1, alone (see Fig. 1.37). From A draw a ray parallel to the axis, which is refracted by L_1 and passes through F_1'. Draw another ray from A passing through O_1, the center of L_1. This ray is not refracted by L_1 and intersects the previous ray at A', forming the image. Draw $A'F_2$ perpendicular to the axis. If F_2 is the focal point of L_2, the rays AA' and $F_1'A'$ will be refracted by L_2 and become parallel to O_2A', which means that the final image will be at infinity.

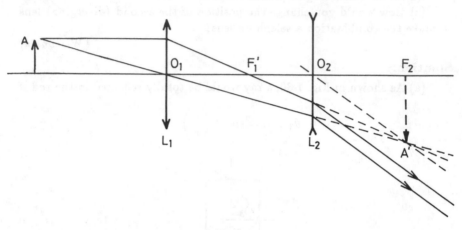

Fig. 1.37

(c) To form a telephoto lens we place the second lens between the first lens and F_1' so that F_1' coincides with F_2. Note that we require $O_2F_2 < O_1F_1'$ (see Fig. 1.38). Rays entering the system with angular spread α will become rays with angular spread α' with $\alpha' > \alpha$ after passing through the optical system. Such a system is commonly known as the Galilean telescope.

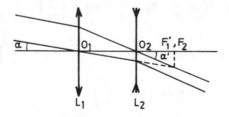

Fig. 1.38

1028

A thin, positive lens L_1 forms a real image of an object located very far way, as shown in Fig. 1.39. The image is located at a distance $4l$ and has height h. A negative lens L_2, of focal length l is placed $2l$ from L_1. A positive lens, of focal length $2l$, is placed $3l$ from L_1, as shown in Fig. 1.40.

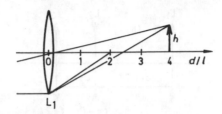

Fig. 1.39

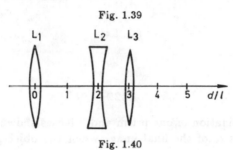

Fig. 1.40

(a) Find the distance from L_1 of the resultant image.

(b) Find the image height.

(*Columbia*)

Solution:

(a) Use the Gaussian lens formula $\frac{1}{u} + \frac{1}{v} = \frac{1}{f}$. Referring to Fig. 1.41 we have

$$L_1 : v_1 = 4l;$$
$$L_2 : f^2 = -l, u_2 = -(4-2)l = -2l, \text{ giving } v_2 = -2l;$$
$$L_3 : f_3 = 2l, u_3 = 2l + l = 3l, \text{ giving } v_3 = 6l.$$

Hence, the distance from L_1 of the resultant image is $3l + 6l = 9l$.

(b) The transverse magnification is $m = \frac{v_2}{u_2} \cdot \frac{v_3}{u_3} = 2$, so the final image height is $h' = mh = 2h$.

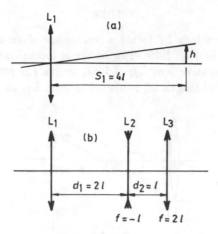

Fig. 1.41

1029

For the combination of one prism and 2 lenses shown (Fig. 1.42), find the location and size of the final image when the object, length 1 cm, is located as shown in the figure.

(*Wisconsin*)

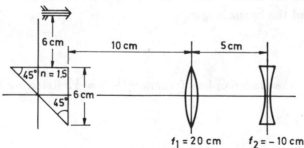

Fig. 1.42

Solution:

For the right-angle prism, $n = 1.5$, the critical angle $\alpha = \sin^{-1}\left(\frac{1}{n}\right) = 42°$, which is smaller than the angle of incidence, $45°$, at the hypotenuse of the prism. Therefore total internal reflection occurs, which forms a virtual image. The prism, equivalent to a glass plate of thickness 6 cm, would cause a image shift of

$$\Delta L = 6\left(1 - \frac{1}{n}\right) = 2 \text{ cm}.$$

Thus the effective object distance for the first lens is

$$u_1 = 10 + 6 + (6 - 2) = 20 \text{ cm}.$$

As $u_1 = f_1$, we have $v_1 = \infty$. Then for the second lens, $u_2 = \infty$. With $f_2 = -10$ cm, we get

$$v_2 = f_2 = -10 \text{ cm}.$$

Hence the final image is inverted and virtual, 10 cm to the left of the second lens.

The size of the image is

$$l_2 = \left|\frac{f_2}{f_1}\right| \cdot 1 = 0.5 \text{ cm}.$$

1030

A 35 mm camera lens of focal length 50 mm is made into a telephoto lens by placing a negative lens between it and the film, as shown (Fig. 1.43). L_1 = camera lens, $f_1 = 50$ mm; L_2 = negative lens, $f_2 = -100$ mm.

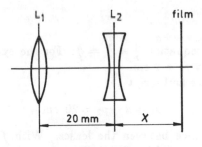

Fig. 1.43

(a) What is the distance x if the system is focused at an object 50 cm in front of L_1?

(b) What is the magnification produced by the lens combination?

(*Wisconsin*)

Solution:

Let $u_1, v_1, f_1, u_2, v_2, f_2$ represent the object distances, image distances and focal lengths for lenses L_1 and L_2, respectively.

(a) $u_1 = 50$ cm $= 500$ mm. The lens formula $\frac{1}{v} = \frac{1}{f} - \frac{1}{u}$ gives $v_1 = \frac{500}{9}$ mm and $u_2 = 20$ mm $-v_1 = -\frac{320}{9}$ mm. It then gives

$$\frac{1}{v_2} = \frac{1}{f_2} - \frac{1}{u_2} = \frac{29}{1600}, \text{ or } x = v_2 = 55.2 \text{ mm}.$$

(b) The magnification of the lens combination is the product of that of each lens:

$$M = \left|\frac{v_1}{u_1}\right| \times \left|\frac{v_2}{u_2}\right| = \frac{\left[\frac{(500\times1600)}{(9\times29)}\right]}{\left[\frac{(500\times320)}{9}\right]}$$

$$= \frac{5}{29} = 0.17.$$

1031

In a compound microscope the focal length of the objective is 0.5 cm and that of the eyepiece is 2 cm. If the distance between lenses is 22 cm, what should the distance from the object to the objective be if the observer focuses for image at ∞? What is the magnifying power? Answer within 10%. Derive all necessary formulae from the lens equation $\frac{1}{p} + \frac{1}{q} = \frac{1}{f}$. Normal near point of eye is 15 cm.

(Wisconsin)

Solution:

Apply the lens equation $\frac{1}{p} + \frac{1}{q} = \frac{1}{f}$. For the eyepiece (subscript 2), $q_2 = \infty$, we get $p_2 = f_2 = 2$ cm.

For the objective (subscript 1),

$$q_1 = d - p_2 = 20 \text{ cm},$$

where d is the distance between the lenses. With $f_1 = 0.5$ cm, we get $p_1 = 0.51$ cm. Hence, the object is located 0.51 cm in front of the objective.

To find the magnifying power, let the object length be y. The length of the image formed by the objective is

$$y' = \frac{q_1}{p_1} y.$$

The angle subtended at the eye by the final image is then

$$\alpha' = \frac{y'}{f_2} = \frac{q_1}{p_1} \frac{y}{f_2} .$$

The angle subtended at the eye by the object without use of microscope is

$$\alpha = \frac{y}{15} .$$

Hence the magnifying power of the microscope is

$$m = \frac{\alpha'}{\alpha} = \frac{15q_1}{p_1 f_2} = 2.9 \times 10^2 .$$

1032

Consider the diagram as shown (Fig. 1.44). A slit source is to be imaged on a screen. The light is to be parallel between the lenses. The index of refraction of the prism is $n(\lambda) = 1.5 + 0.02(\lambda - \lambda_0)/\lambda_0$, where $\lambda_0 = 5000\overset{\circ}{\text{A}}$. System is aligned on 5000 Å light.

(a) What are the focal lengths of the lenses?

(b) What is the linear and angular magnification of the slit at the screen? Is the image inverted? Make a ray diagram.

(c) What is the displacement off axis of light from source of $\lambda = 5050\overset{\circ}{\text{A}}$? Approximate where possible.

(*Wisconsin*)

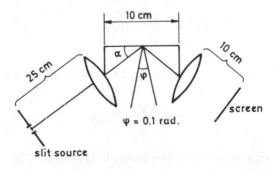

Fig. 1.44

Solution:

(a) As the light emerging from the first lens L_1 is parallel, the slit must be on the front focal plane of L_1, i.e.,

$$f_1 = 25 \text{ cm}.$$

As the parallel light is imaged by the second lens L_2 on the screen, we have

$$f_2 = 10 \text{ cm}.$$

(b) Consider the diagram shown in Fig. 1.45, which represents a co-axial optical system equivalent to the original one. For a prism with small apex angle $(\varphi = 0.1 \text{ rad})$ the angular deviation is very small, $\delta = \varphi(n-1) = 0.05$ rad, and we can consider d, the distance between L_1 and L_2, to be ≈ 10 cm. The positions of the cardinal points of the combination can be calculated as follows,

$$\Delta = d - f_1 - f_2 = -25 \text{ cm},$$
$$f = -\frac{f_1 f_2}{\Delta} = 10 \text{ cm},$$
$$\overline{A_1 H} = \frac{fd}{f^2} = 10 \text{ cm},$$
$$\overline{A_2 H'} = -\frac{fd}{f_1} = -4 \text{ cm},$$
$$\overline{F_1 F} = \frac{f_1^2}{\Delta} = -25 \text{ cm},$$
$$\overline{F_2' F'} = -\frac{f_2^2}{\Delta} = 4 \text{ cm}.$$

Thus F' is 4 cm in front of the screen, i.e., $x' = 4$ cm, so the linear magnification is

$$m = -\frac{x'}{f} = -0.4.$$

The negative sign shows that the image is inverted. The ray diagram is shown in Fig. 1.45.

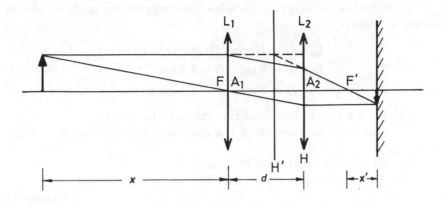

Fig. 1.45

The angular magnification, which is the ratio of the tangents of the slopes of conjugate rays to the axis, is given by

$$M = \frac{-1}{f + x'} \left(\frac{1}{f + x} \right)^{-1} = -2.5 \, .$$

(c) The angular derivation δ is a function of wavelength λ through

$$\delta - [n(\lambda) - 1]\varphi \, .$$

Hence

$$\Delta\delta = \varphi \frac{dn}{d\lambda} \Delta\lambda = 0.1 \times \frac{0.02}{5000} \times (5050 - 5000) = 2 \times 10^{-5} \, .$$

1033

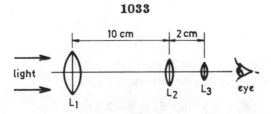

Fig. 1.46

Consider the following simple telescope configuration made up of thin lenses as follows:

$$L_1 : f_1 = 10 \text{ cm}, \text{ Diameter} = 4 \text{ cm}$$
$$L_2 : f_2 = 2 \text{ cm}, D = 1.2 \text{ cm}$$
$$L_3 : f_3 = 2 \text{ cm}, D = 1.2 \text{ cm}$$

(a) Trace a typical bundle of rays through the system.

(b) Calculate the position of the exit pupil and the diameter of the exit pupil.

(c) What is the function of the lens L_2?

(d) Is the instrument a good match to the eye? (Explain).

(*Wisconsin*)

Solution:

(a) Let us consider a bundle of parallel rays falling normally on L_1 as shown in Fig. 1.47. We are given that the back focal point of L_1 and the front focal point of L_3 coincide with the center of L_2, so the bundle of parallel rays suffers no refraction at L_2.

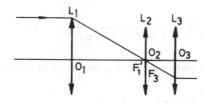

Fig. 1.47

(b) The lens L_1 functions as the aperture and the entrance pupil of the telescope. The image of the aperture of L_1 formed by the following lenses is the exit pupil of the telescope, the position of which can be calculated using the lens formula twice:

$$\frac{1}{S_2} + \frac{1}{S_2'} = \frac{1}{f_2},$$
$$\frac{1}{S_3} + \frac{1}{S_3'} = \frac{1}{f_3},$$

where $S_2 = 10$ cm, $f_2 = f_3 = 2$ cm.

Solving the two equations in turn, we have

$$S_2' = 2.5 \text{ cm}, \quad S_3 = 2 - 2.5 = -0.5 \text{ cm},$$

and hence

$$S_3' = 0.4 \text{ cm},$$

i.e., the exit pupil is located 0.4 cm behind L_3. From the similar triangles we have $\frac{D'}{D} = \frac{f_3}{f_1}$, so that the diameter of the exit pupil is

$$D' = \frac{f_3}{f_1} D = 0.8 \text{ cm}.$$

(c) The lens L_2 placed at the front focal point of L_3 is called the field lens and serves as the field stop in the configuration. Its function is to converge the rays before they enter the eye lens L_3, so that one can get a comfortable image with a smaller eye lens. Furthermore, the use of a system of L_2 and L_3 of the same glass separated by $(f_2 + f_3)/2$, which has the same focal length for all colors, as the ocular eliminates lateral chromatic observation.

(d) In addition to having a large aperture objective to give the necessary resolving power, the objects will only be resolved if the magnification is sufficiently large to present the necessary separation to the eye. For this purpose the minimum magnification required is

$$m_0 = \frac{00}{\frac{140}{D}},$$

where D is in centimeters. However, for comfort of the eye the actual magnification should be 1.5 to 2 times the minimum, i.e.,

$$m = \frac{120 D}{140} = 3.4$$

with $D = 4$ cm as given.

The magnification for the given system is

$$\frac{f_1}{f_3} = 5 > m.$$

Hence, as far as magnification is concerned the system is quite up to the standard. On the other hand, for a good optical system, the exit pupil should be designed to coincide with the eye and the distance between the exit pupil and the last optical surface of the eye lens should not be less than 5 mm. Such distance for this system is only 4 mm, which would make the eyelash touch the optical surface. Moreover, the size of the exit pupil of a telescope should be as large as the eye, i.e., 2–4 mm. The diameter of the exit pupil of this system is 8 mm, so the light emerging from the telescope will not all enter the eye.

For these reasons, the telescope is not a good match to the eye.

1034

Two telescopes have the same objective lens with focal length f_0. One telescope has a converging lens for an eyepiece with focal length f_c. The other has a diverging lens for an eyepiece with focal length $-f_d$. The magnification for these two telescopes is the same for objects at infinity. What is the ratio of the lengths of these two telescopes in terms of the magnification M? Give a good reason why one might choose to use the longer telescope for a given M.

(*Wisconsin*)

Solution:

The magnification, usually the angular magnification, of a visual instrument is defined as the ratio of the angles ω' and ω which are made by the chief rays from the top of the object with the axis at the aided and unaided eyes respectively:

$$M = \frac{\omega'}{\omega}.$$

For a telescope

$$M = -\frac{f_0}{f_e}.$$

where f_0 and f_e are the focal lengths of the objective lens and the eyepiece respectively. A telescope system uses a converging or a diverging lens as the eyepiece. A telescope with converging eyepiece is called a Keplerian telescope, while that with diverging eyepiece is called a Galilean telescope. For the Keplerian and Galilean telescopes given the magnifications are

$$M_K = -\frac{f_0}{f_c}$$

and

$$M_G = \frac{-f_0}{-f_d} = \frac{f_0}{f_d}$$

respectively, where the minus sign corresponds to an inverted image and the plus an erect one. We are given that

$$-M_K = M_G = M,$$

hence

$$f_c = f_d \, .$$

For a telescope, the secondary focal point of the objective coincides with the primary focal point of the eyepiece. Thus the length of a telescope is $L = f_0 + f_e$.

For the two telescopes given, we have

$$L_K = f_0 + f_c \, , \qquad L_G = f_0 - f_d \, .$$

Obviously,

$$L_K > L_G \, ,$$

$$\frac{L_K}{L_G} = \frac{(f_0 + f_c)}{(f_0 - f_d)} = \frac{((f_0/f_c) + 1)}{((f_0/f_d) - 1)} = \frac{(M + 1)}{(M - 1)} \, .$$

Comparing with the Galilean telescope, the advantages of the Keplerian telescope are as follows:

(a) Between the objective lens and the eyepiece, there is a real image plane where a graticule can be placed as reference mark for making measurements.

(b) A field stop can be placed in the image plane, so that the entrance window overlaps the object plane at infinity. This makes a higher quality image with a larger field of view and no vignetting.

However, the Keplerian telescope has the disadvantage that it forms an inverted image, which is inconvenient for the observer. Adding an image-inverter could eliminate this defect. Therefore, although the length of a Keplerian telescope is longer than that of a Galilean one for the same magnification, the former is still to be preferred to the latter.

1035

(a) Derive the lens makers' formula for a thin lens

$$\frac{1}{f} = (n - 1) \left(\frac{1}{R} + \frac{1}{R'} \right) \, .$$

(b) Crown and flint glass lenses are cemented together to form an achromatic lens. Show that the focal lengths of the crown and flint components satisfy the equation

$$\frac{\Delta_c}{f_c} + \frac{\Delta_f}{f_f} = 0 \, ,$$

where

$$\Delta = \frac{1}{\bar{n}-1} \frac{dn}{d\lambda} \Delta\lambda$$

and $\bar{n}$ is the value of the index of refraction of the glass at the center of the visible spectrum and $\Delta\lambda$ the wavelength spread ($\sim 3000\overset{\circ}{\text{A}}$) of the visible spectrum.

(c) For crown glass $\Delta_c = 0.0169$ and for flint glass $\Delta_f = 0.0384$. Show that one must use a converging crown glass lens cemented to a diverging flint glass lens to get a converging achromatic lens.

(Columbia)

Solution:

(a) Please refer to any standard textbook.

(b) For the compound lens formed by cementing the two component lenses together, the focal length f satisfies the equation

$$\frac{1}{f} = \frac{1}{f_c} + \frac{1}{f_f}. \tag{1}$$

Taking derivatives with respect to λ yields

$$\frac{1}{f^2} \frac{df}{d\lambda} = \frac{1}{f_c^2} \left(\frac{df_c}{d\lambda}\right) + \frac{1}{f_f^2} \left(\frac{df_f}{d\lambda}\right). \tag{2}$$

Differentiating the lens makers' formula

$$\frac{1}{f} = (n-1)\left(\frac{1}{R} + \frac{1}{R'}\right), \tag{3}$$

we have

$$\frac{1}{f^2} \frac{df}{d\lambda} = \left(\frac{1}{R} + \frac{1}{R'}\right) \frac{dn}{d\lambda},$$

or, making use of (3),

$$\frac{1}{f} \frac{df}{d\lambda} = \frac{1}{\bar{n}-1} \frac{dn}{d\lambda} = \frac{\Delta}{\Delta\lambda},$$

or,

$$\frac{1}{f^2} \frac{df}{d\lambda} = \frac{\Delta}{\Delta\lambda} \frac{1}{f}.$$

So we have

$$\frac{1}{f_c^2} \frac{df_c}{d\lambda} = \frac{\Delta_c}{\Delta\lambda} \frac{1}{f_c}$$

and

$$\frac{1}{f_f^2}\frac{df_f}{d\lambda} = \frac{\Delta_f}{\Delta\lambda}\frac{1}{f_f}.$$

Substituting these expressions into (2) and noting that $\frac{df}{d\lambda} = 0$ for an achromatic lens, we have

$$\frac{\Delta_c}{f_c} + \frac{\Delta_f}{f_f} = 0. \tag{4}$$

(c) Solving (1) and (4), we have

$$\frac{1}{f_c} = \frac{\Delta_f}{f(\Delta_f - \Delta_c)},$$

$$\frac{1}{f_f} = \frac{-\Delta_c}{f(\Delta_f - \Delta_c)}.$$

Given that $\Delta_c < \Delta_f$ and the requirement $f > 0$, we have

$$f_c > 0 \quad \text{and} \quad f_f < 0.$$

Therefore, to get a converging achromatic lens one must use a converging crown glass lens cemented to a diverging flint glass lens.

1036

A common piece of optical equipment is the viewgraph machine used in physics colloquia and seminars. A rough sketch is shown below (Fig. 1.48).

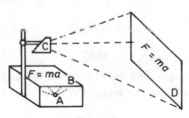

Fig. 1.48

One or more optical elements are located at positions A, B, C, and D. Describe each element and explain how the elements are used in the operation of the viewgraph machine. Write down the relations which must be satisfied by the distances between the elements. Which element in this device was not available at reasonable cost in 1900?

(*Princeton*)

Solution:

There are two types of viewgraph machines, transmission and reflection. The equipment shown belongs to the former type.

A is the illumination system and consists of a lamp and a couple of condensing lenses. B is a transparency to be projected. C is the image formation and projection system consisting of an objective lens and a mirror. D is the screen.

The light, emitted by the lamp and condensed by the condensing lenses, illuminates the "object" – the transpareney, on which there are words or drawings. The object is imaged by the objective lens and then reflected and projected by the mirror onto the screen to form an enlarged image.

The objective lens is the key element, upon which the quality of the resultant image depends. The objective lens has to be specially designed and manufactured in order to eliminate, as much as possible, spherical and chromatic aberrations, astigmatism and field curvature. Sometimes one even employs a camera objective lens for this purpose. So the objective must have been very expensive in 1900.

The image plane of the source formed by the condensing lenses should coincide with the pupil of the objective lens. That is to say, the lamp and the pupil of the objective lens are conjugate for the condenser. The transparency and the screen should also be a pair of conjugate object and image for the objective lens. The distance between the transparency and the objective lens, the distance between the objective lens and the screen (including the distance from the objective lens to the center of the mirror and that from the center of the mirror to the screen) and the focal length of the objective lens should satisfy the lens makers' formula.

1037

A lens (of focal length f) gives the image of the sun on its focal plane. Prove that the brightness of the image (W/cm^2) is near that of the surface of the sun.

(Columbia)

Solution:

The sun may be considered a Lambert source with apparent luminance, or brightness, L and area S. The power radiated into the lens (of area A and focal length f) is

$$\Phi = \frac{LSA}{R^2} ,$$

where R is the distance between the sun and the earth. All of the power collected is directed through S' (the image of S).

The apparent luminance L' of the image is

$$\Phi' = L'S'd\Omega',$$

where $d\Omega'$ is the solid angle subtended by the lens at the image, i.e.,

$$d\Omega' = \frac{A}{f^2}.$$

If we neglect attenuation in the atmosphere and the lens, then $\Phi \approx \Phi'$, i.e.,

$$\frac{LSA}{R^2} \approx \frac{L'AS'}{f^2}.$$

As $S/R^2 = S'/f^2$, we have

$$L \approx L'.$$

1038

(It is impossible to increase the apparent brightness of an extended (large solid angle) diffuse light source with lenses. This problem illustrates that fact for a single lens.) A light source of brightness S subtends a solid angle that is larger than the acceptance solid angle Ω of a telescope that observes it. The source emits S units of optical energy per unit area per unit solid angle per second isotropically. The objective lens of the telescope has area A and is a thin lens.

(a) Show that the energy entering the telescope per second is $S\Omega A$.

(b) Show that the product of the area of the image formed by the objective lens and the solid angle subtended by the objective lens at the image is ΩA.

(c) Explain why the above results show that the apparent brightness of the extended source is not changed by the objective lens of the telescope.

(*UC, Berkeley*)

Solution:

(a) The light flux entering the telescope per second is given by

$$\phi = \int (S d\Sigma \ d\Sigma' \ \cos\theta \cos'/r^2),$$

where $d\Sigma$ is an area element of the source surface and $d\Sigma'$ is an area element of the telescope lens, θ is the angle which the telescope axis makes with the normal to $d\Sigma$, θ' is the angle which the axis makes with the normal to $d\Sigma'$, r is the distance between $d\Sigma$ and $d\Sigma'$. As $\int(d\Sigma/r^2)$, the solid angle subtended by the source, is larger than the acceptance solid angle Ω of the telescope, only light emitted from an effective area Ωr^2 enters into the telescope, i.e., $\int(d\Sigma/r^2) = \Omega$. Then as $\int d\Sigma' = A$, we have

$$\Phi = S\Omega A.$$

Here we have assumed that the distance between the source and the telescope is very large and that the telescope observes directly, so that both $\cos\theta$ and $\cos\theta'$ are each equal to unity.

(b) The area of the image, σ', is given by

$$\sigma' = \Omega f^2,$$

where f is the focal length of the objective lens of the telescope. The solid angle Ω' subtended by the lens at the image is

$$\Omega' = \left(\frac{\pi D^2}{4}\right) \bigg/ f^2,$$

where D is the diameter of the lens. Therefore we have

$$\sigma'\Omega' = \frac{\pi D^2}{4f^2} \cdot \Omega f^2 = \frac{\pi D^2}{4} \cdot \Omega = A \cdot \Omega.$$

(c) The light flux in the image area is

$$\Phi' = \pi S'\sigma' \sin^2 u',$$

where S' is the brightness of the image, u' is the angular semi-aperture of the lens, given by $\sin u' = \frac{D}{2f}$. Hence

$$\Phi' = \pi S'(\Omega f^2)\left(\frac{D}{2f}\right)^2 = S'\Omega A.$$

As the transmission coefficient of the lens is smaller than unity, i.e., $\Phi' \leq \Phi$, we get $S' \leq S$. That is, it is impossible to increase the apparent brightness of an extended diffuse source with a telescope.

1039

When the sun is overhead, a flat white surface has a certain luminous flux. A lens of radius r and focal length f is now used to focus the sun's image on the sheet. How much greater is the flux in the image area? For a given r, what must f be so that the lens gives no increase in flux in the image? From the earth the diameter of the sun subtends about 0.01 radians. The only light in the image is through the lens.

(UC, Berkeley)

Solution:

The luminous flux reaching the earth is given by

$$\Phi = B\sigma' d\Omega$$

where B is the brightness of the sun, regarded as a Lambert radiator, σ' is the area illuminated on the earth, $d\Omega$ is the solid angle subtended by the sun at σ' and is given by

$$d\Omega = \pi\alpha^2,$$

α being the angular aperture of the sun, equal to about 0.01 radians.

The luminous flux on the image area behind the lens is

$$\Phi' = \pi B\sigma' \sin^2 u' = \pi B\sigma' \left(\frac{r}{f}\right)^2$$

where u' is the semi-angular aperture of the lens, r and f are the radius and focal length of the lens respectively. Consider the ratio

$$\frac{\Phi'}{\Phi} = \frac{\pi B\sigma'\left(\frac{r^2}{f^2}\right)}{\pi B\sigma'(0.01)^2} = \frac{10^4 r^2}{f^2},$$

where we have taken the transparency of the lens glass to be unity. Thus the luminous flux on the image area is $10^4 r^2/f^2$ times that on a surface of the same area on the earth without the lens.

For $\Phi'/\Phi \leq 1$, we have $f \geq 10^2 r$. Hence, for $f \geq 100r$ the lens will give no increase in flux in the image area.

1040

(a) Three identical positive lenses, of focal length f, are aligned and separated by a distance f from each other, as shown in Fig. 1.49. An object is located $f/2$ in front of the leftmost lens. Find the position and the magnification power of the resultant image by using tracing method.

(b) Looking through a small hole is a well-known method to improve sight. If your eyes are near-sighted and can focus an object 20 cm away without using any glasses, estimate the required diameter of the hole through which you would have good sight for objects far away.

(Columbia)

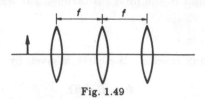

Fig. 1.49

Solution:

(a) We construct the image using two rays passing through the top of the object, one parallel to the axis, the other through the front focal point of L_1, as shown in Fig. 1.50. The resultant image is found at a distance $f/2$ behind the rightmost lens. The magnification is -1.

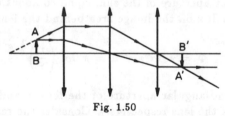

Fig. 1.50

(b) Inside the human eye the distance between the crystalline lens and the retina is 20 mm. Using the lens formula $\frac{1}{u} + \frac{1}{v} = \frac{1}{f}$, we get $f = 19$ cm for the image distance 2.0 cm and the object 20 cm away.

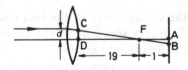

Fig. 1.51

As Fig. 1.51 shows, a lens focuses an object infinitely far away at the focal point F and forms a disk AB on the retina 1.0 cm behind F. An opening placed in front of the eye lens determines the diameter of the beam of light entering the eye. From the geometry, $d = \overline{CD} = 19\,\overline{AB}$. On account of the resolving angular limitation of the eye, which is about $1' = 3 \times 10^{-4}$ rad, we get a transverse resolving limitation of $\overline{AB}_m = 3 \times 10^{-4} \times 20 = 6 \times 10^{-3}$

mm. If the diameter of the opening $d < 19 \times 6 \times 10^{-3} = 0.12$ mm, we would have good sight for objects far away.

1041

One side of a disk, 1 cm^2 area, radiates uniformly and isotropically (a Lambert's law radiator) with a brightness of 1 watt $\cdot$ cm^{-2} steradian^{-1} at a single frequency in the visible.

(a) What is the total rate at which energy is radiated from this face of the disk?

(b) Given a fused quartz lens ($n = 1.5$) whose diameter is 10 cm and whose focal length is 100 cm, show how to image the radiator onto a disk whose area is $\frac{1}{4}$ cm^2.

(c) Estimate the total energy flux reaching the $\frac{1}{4}$ cm^2 disk to within a few percent.

(d) By changing n and the dimensions of the lens, you may increase the energy flux reaching the $\frac{1}{4}$ cm^2 disk. By what reasoning might you determine the maximum energy flux into the $\frac{1}{4}$ cm^2 disk that can be achieved?

(*CUSPEA*)

Solution:

(a) $\Phi = BS \int_0^{2\pi} d\varphi \int_0^{\pi} \sin\theta \cos\theta d\theta = \pi BS = 3.14$ W,

where B is the brightness of the source, S the area of the source, θ the angle between the direction of the emitted light and the normal to the surface of the source.

(b) The magnification is given by

$$m = \frac{D'}{D} = \left(\frac{S'}{S}\right)^{\frac{1}{2}} = 0.5,$$

where D', D, S' and S are the diameters and the areas of the image and the object respectively. As $m = \frac{v}{u} = 0.5$, we have $u = 2v$.

The lens equation then gives $u = 3f = 300$ cm, $v = \frac{3f}{2} = 150$ cm. Therefore, if the object and image distances are 300 cm and 150 cm respectively, the lens images the source onto a disk of area 0.25 cm.

(c) If the lens does not cause attenuation, the energy flux reaching the $\frac{1}{4}$ cm^2 disk would be equal to that reaching the lens. The solid angle subtended by the lens at the source is

$$\Omega = \frac{\pi r^2}{u^2}.$$

The total energy flux reaching the lens is then

$$\Phi' = BS\Omega = 1 \times 1 \times \pi \frac{5^2}{300^2} = 8.7 \times 10^{-4}\,\text{W}.$$

(d) According to the second law of thermodynamics, the brightness of the image cannot exceed that of the object. So the maximum energy flux into the $\frac{1}{4}$ cm^2 disk is

$$\Phi'' = \frac{\Phi S'}{S} = \frac{\Phi}{4} = 0.785 \quad \text{W}.$$

PART 2 WAVE OPTICS

The width of a certain spectral line at 500 nm is 2×10^{-2} nm. Approximately what is the largest path difference for which interference fringes produced by this light are clearly visible?

(Wisconsin)

Solution:

The coherence length l_c is given by

$$.\lambda_c = \frac{\lambda^2}{\Delta\lambda} = 1.25 \times 10^{-3} \text{ cm} .$$

If the optical path difference is about a quarter of l_c, 3×10^{-4} cm, we can observe the fringes clearly.

A point source S located at the origin of a coordinate system emits a spherical sinusoidal wave in which the electric field E_1 is given by $E_1 = A(\frac{D}{r})\cos(\omega t - \frac{2\pi r}{\lambda})$, where r is the distance from S. In addition, there is a plane wave propagating along the x-axis. This wave is given by

$$E_2 = A \cos\left(\omega t - \frac{2\pi x}{\lambda}\right) .$$

(Note that we treat both E_1 and E_2 as scalar waves in this problem.) Both waves are incident on a flat screen perpendicular to the x-axis and at a distance D from the origin, as shown in the figure (Fig. 2.1). Compute the resultant intensity I at the screen as a function of the distance y from the x-axis for values of y small compared with D. Express I in terms of y, D, λ and the intensity I_0 at $y = 0$.

(Wisconsin)

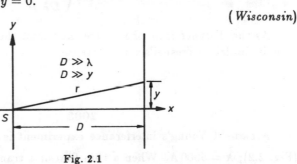

Fig. 2.1

Solution:

The electric field E at a point, distance r from the origin, on the screen is given by

$$E = E_1 + E_2 = A \left[\frac{D}{r} \cos \left(\omega t - \frac{2\pi r}{\lambda} \right) + \cos \left(\omega t - \frac{2\pi x}{\lambda} \right) \right]$$

$$\approx A \left[\cos \left(\omega t - \frac{2\pi r}{\lambda} \right) + \cos \left(\omega t - \frac{2\pi x}{\lambda} \right) \right]$$

where we have used the approximation $r \approx D$.

For $y \ll D$, we have

$$r = \left(D^2 + y^2 \right)^{\frac{1}{2}} \approx D \left(1 + \frac{y^2}{2D^2} \right)$$

and thus

$$E = A \left[\cos \left(\omega t - \frac{2\pi D \left(1 + \frac{y^2}{2D^2} \right)}{\lambda} \right) + \cos \left(\omega t - \frac{2\pi D}{\lambda} \right) \right]$$

$$= 2A \cos \frac{\pi y^2}{D\lambda} \cdot \cos \left(\omega t - \frac{2\pi D \left(1 + \frac{y^2}{4D^2} \right)}{\lambda} \right).$$

Hence

$$I \propto E^2 \propto \cos^2 \frac{\pi y^2}{D\lambda}.$$

Let $I = I_0$ at $y = 0$; we then have

$$I = I_0 \cos^2 \left(\frac{\pi y^2}{D\lambda} \right).$$

As the distance from the centre increases, the fringes become closer. This is similar to Fresnel's zone plate.

2003

A two-slit Young's interference experiment is arranged as illustrated (Fig. 2.2); $\lambda = 5000 \overset{\circ}{A}$. When a thin film of a transparent material is put behind one of the slits, the zero order fringe moves to the position previously occupied by the 4th order bright fringe. The index of refraction of the film is $n = 1.2$. Calculate the thickness of the film.

(*Wisconsin*)

Solution:

Intensity maxima occur when the optical path difference is $\Delta = m\lambda$. Thus

$$\delta\Delta = \lambda\delta m.$$

When the film is inserted as shown in Fig. 2.2, the optical path changes by

$$\delta\Delta = t(n-1),$$

where t is the thickness of the film. As the interference pattern shifts by 4 fringes,

$$\delta m = 4.$$

Hence $t(n-1) = 4\lambda$, giving

$$t = \frac{4\lambda}{n-1} = 10 \quad \mu m.$$

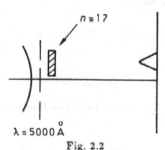

Fig. 2.2

2004

Consider the interference pattern from the three slits illustrated (Fig. 2.3). Assume the openings of the individual slits are the same $(\lesssim\frac{\lambda}{2})$.

(a) At what value of θ is the first principal maximum? (i.e., the wavelets from all three slits are in phase).

(b) The result for (a) can be called θ_1. The flux in the direction of zeroth order maximum $(\theta = 0)$ is F_0. What is the flux (in units of F_0) in the direction $\frac{\theta_1}{2}$? (Assume $\lambda \ll d$.)

(*Wisconsin*)

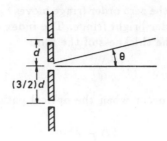

Fig. 2.3

Solution:

(a) The electric field $E(\theta)$ on the screen is the sum of the fields produced by the slits individually:

$$E(\theta) = E_1 + E_2 + E_3 = A + Ae^{i\delta} + Ae^{i\frac{5}{2}\delta},$$

where

$$\delta = \frac{2\pi d}{\lambda} \sin \theta.$$

Then the total intensity at θ is

$$I(\theta) \sim E(\theta) \cdot E^*(\theta) = A^2 \left\{ 3 + 2 \left[\cos \delta + \cos \left(\frac{3\delta}{2} \right) + \cos \left(\frac{5\delta}{2} \right) \right] \right\}.$$

For $\theta = 0$, we have

$$I(0) \sim 9A^2.$$

The expression for $I(\theta)$ shows that the first principal maximum occurs at $\delta = 4\pi$, i.e.,

$$\left(\frac{2\pi d}{\lambda} \right) \sin \theta_1 = 4\pi,$$

or

$$\theta_1 \approx \sin \theta_1 \sim \frac{2\lambda}{d}.$$

(b)

$$I\left(\frac{\theta_1}{2} \right) \sim A^2 [3 + 2(\cos 2\pi + \cos 3\pi + \cos 5\pi)] = A^2 \sim \frac{I(0)}{9}.$$

2005

An opaque screen with two parallel slits is placed in front of a lens and illuminated with light from a distant point source.

(a) Sketch the interference pattern produced. (Make a rough graph of the intensity distribution in the focal plane of the lens.)

(b) Indicate with a second graph and a brief explanation the effect of moving the slits further apart.

(c) Indicate with a third graph and explanation the effect of increasing the size of the source so that it subtends a finite angle at the lens.

(*Wisconsin*)

Solution:

(a) The interference picture is shown in Fig. 2.4. The spacing of fringes is $\Delta x = \frac{\lambda f}{d}$, where d is the distance between the two slits, f the focal length of the lens.

(b) As d increases, Δx decreases, fringes become denser. The interference pattern is shown in Fig. 2.5.

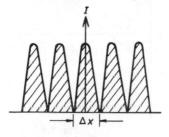

Fig. 2.4

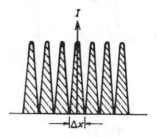

Fig. 2.5

(c) Superposition of all incoherent, double-slit interference patterns produced by individual parts of the source decreases the contrast of the resultant pattern, as shown in Fig. 2.6.

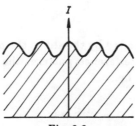

Fig. 2.6

2006

Consider the Young's interference experiment shown in the diagram (Fig. 2.7).

Assume that the wavelength of the light is $6000\,\overset{\circ}{A}$, the slit widths are all the same, $S_0 = S_1 = S_2 = 0.2$ mm, the slit separation $d = 2.0$ mm, and $L_1 = 3.0$ m.

(a) About how long must L_0 be in order to produce a well-defined interference pattern on the screen?

(b) What is the distance between the central and first bright fringe on the screen?

(Wisconsin)

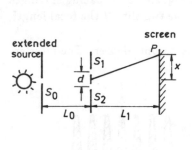

Fig. 2.7   Fig. 2.8

Solution:

(a) According to the Van Cittert Zernike theorem, the lateral coherence length of an extended slit source is given by

$$I_s = \frac{\lambda}{\theta_s},$$

where θ_s is the angle subtended by the slit at the point concerned in radians. In order to get a clear interference pattern, the distance between S_1 and S_2 should be less than l_s, i.e.,

$$d < l_s,$$

or

$$\theta_s < \frac{\lambda}{d}.$$

As

$$\theta_s = \frac{S_0}{L_0},$$

we require that

$$L_0 > \frac{d}{\lambda}S_0 = \frac{2 \times 10^{-4}}{6000 \times 10^{-10}} \times 2 \times 10^{-3} = 0.67 \quad \text{m}.$$

(b) Referring to Fig. 2.8, consider the sum of the contributions to the amplitude at P of the screen of width elements dy at the points y and $-y$ of the slits S_1 and S_2 respectively. As $d \gg S_1, S_2$, it is proportional to

$$\left[e^{ik(x-y\sin\theta)} + e^{ik(x+y\sin\theta)}\right] dy = e^{ikx}\cos(ky\sin\theta)dy,$$

where $k = \frac{2\pi}{\lambda}$.

Then the total amplitude at P is

$$A \propto \int_{\frac{d}{2}-\frac{S}{2}}^{\frac{d}{2}+\frac{S}{2}} \cos(ky\sin\theta)dy$$

as $S_1 = S_2 = S$. With $\sin\theta \approx \frac{x}{L_1}$, we have

$$A(x) = 2A_0 S \left(\frac{\sin\frac{\pi x}{\lambda L_1}S}{\frac{\pi x}{\lambda L_1}S}\right) \cdot \cos\left(\frac{\pi x}{\lambda L_1}d\right),$$

where $A_0 = $ constant. The intensity $I(x)$ is therefore

$$I(x) \propto \cos^2\left(\frac{\pi x}{\lambda L_1}d\right).$$

For $\frac{\pi x d}{\lambda L_1} = \pi$, the first maximum occurs, corresponding to

$$x = \frac{\lambda L_1}{d} = 0.9 \quad \text{mm}.$$

Therefore, the first bright fringe is 0.9 mm away from the center of the interference pattern.

2007

A two-slit diffraction pattern is produced by the arrangement shown in Fig. 2.9. A discharge tube produces light of wavelength λ which passes through a small slit S immediately in front of the tube. The centers of the two slits of width w are at distance D apart. Find the condition that a maximum intensity be observed at the screen a distance l from the central plane. Assume that $D \ll L, l \ll L$, and that w is very small. How large may w be before the interference pattern is washed out? As w increases, do the maxima and minima wash out first near the midplane or further away? What is the effect on the intensity and sharpness of the pattern of increasing the width S of the slit at the source?

(*Wisconsin*)

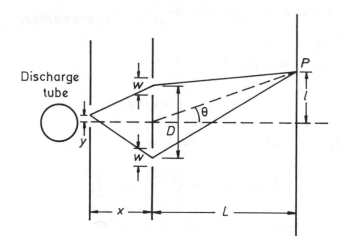

Fig. 2.9

Solution:

The intensity distribution on the screen produced by the elemental slit of width dy of the slit sources is given by

$$dI = 2I_0 \left(\frac{\sin \beta}{\beta} \right)^2 (1 + \cos \delta) dy,$$

where $\beta = \frac{\pi w l}{\lambda L} = \frac{\pi w \sin \theta}{\lambda}$ with $\sin \theta = \frac{l}{L}, \delta = \delta_1 + \delta_2$ with

$$\delta_1 = 2\pi D \sin \theta / \lambda, \quad \delta_2 = 2\pi D y / \lambda x,$$

and I_0 is a constant. Integration yields the intensity distribution produced by the slit source:

$$I = 2I_0 \left(\frac{\sin \beta}{\beta} \right)^2 \int_{-\frac{s}{2}}^{\frac{s}{2}} (1 + \cos \delta) dy$$

$$= 2I_0 S \left(\frac{\sin \beta}{\beta} \right)^2 \left\{ 1 + \frac{\sin \left(\frac{\pi D S}{\lambda x} \right)}{\left(\frac{\pi D S}{\lambda x} \right)} \cos \left(\frac{2\pi D}{\lambda} \sin \theta \right) \right\} .$$

If $2\pi D \sin \theta / \lambda = 2\pi Dl/\lambda L = 2n\pi$, i.e., $l = \frac{n\lambda L}{D}$, where n is an integer, maximum intensity is observed on the screen at distance l from the central plane.

The interference pattern is washed out for the first time if $\sin \beta = 0$, or $\beta = \pi w l/(\lambda L) = \pi$, i.e., $w = \frac{\lambda L}{l} = D$. In general, $w = \frac{D}{n}$, or $n = \frac{D}{w}$. As w increases, n decreases. That is to say, the maxima and minima will first wash out further away from the midplane.

The visibility of the fringes is determined by

$$V = \left(\frac{I_{\max} - I_{\min}}{I_{\max} + I_{\min}} \right) = \left| \frac{\sin \frac{\pi DS}{\lambda x}}{\frac{\pi DS}{\lambda x}} \right| ,$$

where $I_{\max}$ and $I_{\min}$ are the intensities at the maxima and minima of the fringe pattern. If $\pi DS/(\lambda x) = n\pi$, where n is an integer, then $V = 0$. Within $S_0 = \lambda x/D$, as S increases, the visibility of the fringes decreases.

From the expression for I, it is obvious that the intensity on the screen is proportional to the width S of the slit at the source.

2008

Fresnel biprism. A Fresnel biprism of refractive index n and small equal base angles α is constructed as shown in the sketch (Fig. 2.10).

(a) A ray of light incident from the left normal to the base of the prism may enter either the upper or lower half. Obtain the angular deviation θ for the two cases. Assume α small. Draw and label a sketch.

(b) A plane wave is incident normal to the base of the prism, illuminating the entire prism. Fringes are observed in the transmitted light falling on a screen parallel to the base of the prism. What is the origin of these fringes? Obtain an expression for the fringe separation in terms of the angle of deviation θ of an incident ray. Draw and label a sketch.

(c) With a glass biprism in yellow light a fringe separation of 100 μm (10^{-2} cm) is observed. Estimate in degrees the base angles α of the prism. Indicate and justify your choice of refractive index and wavelength of yellow light.

(*UC, Berkeley*)

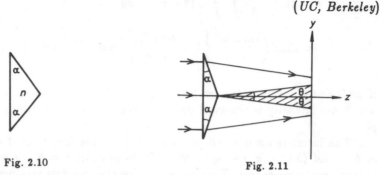

Fig. 2.10   Fig. 2.11

Solution:

(a) A beam of light incident normal on the base undergoes refraction at the side surfaces. A ray of light emerging from the upper half of the prism travels obliquely downwards, and by symmetry, one from the lower half travels obliquely upwards, as shown in Fig. 2.11. The magnitudes of the angular deviation for the two cases are the same, being given for small α by

$$\theta = (n-1)\alpha .$$

(b) The two parallel beams of light emerging symmetrically from the upper and lower halves of the prism meet and interfere with each other and interference fringes along the x axis appear on the screen. The spacing of the fringes is given by

$$\Delta y = \frac{\lambda}{2\sin\theta} \approx \frac{\lambda}{2\theta} = \frac{\lambda}{2(n-1)\alpha} .$$

(c) For yellow light of $\lambda = 6000 \overset{\circ}{A}$ and biprism of index of refraction $n = 1.5$, $\Delta y = 100\mu m$ gives $\alpha = 6 \times 10^{-3}\,\mathrm{rad} = 21'$.

2009

A Lloyd's mirror (see Fig. 2.12) can be used to obtain interference fringes on the screen from a single source as shown. Construct (draw) a two (coherent) source system which is equivalent. The first fringe (at the point 0) is dark. What does this imply?

(Wisconsin)

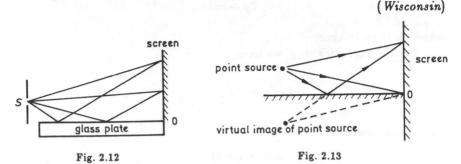

Fig. 2.12   Fig. 2.13

Solution:

A two-source system equivalent to the Lloyd's mirror is shown in Fig. 2.13. The dark fringe at the point 0 implies that the beam reflected from the mirror surface undergoes a phase change of π on reflection.

2010

A Michelson interferometer is adjusted to give a fringe pattern of concentric circles when illuminated by an extended source of light of $\lambda = 5000\overset{\circ}{A}$. How far must the movable arm be displaced for 1000 fringes to emerge from the center of the bullseye? If the center is bright, calculate the angular radius of the first dark ring in terms of the path length difference between the two arms and the wavelength λ.

(Wisconsin)

Solution:

Concentric circles, called fringes of equal inclination, are obtained when the two reflecting mirrors are exactly mutually perpendicular. The optical path length difference Δ between the two arms is given by

$$\Delta = 2nd\cos\theta = 2n(l_1 - l_2)\,\cos\theta\,,$$

where n is the index of refraction of the medium (for air, $n = 1$), $d = l_1 - l_2$ is the path length difference (OPD) between the arms of lengths l_1 and l_2, θ is the angle of incidence of light at the mirrors.

(1) For 1000 fringes to emerge from the center of the bullseye, the OPD Δ must undergo a change of 1000λ. Thus $2d = 1000\lambda$, or $d = 500\lambda = 0.25$ mm, i.e., the movable arm is displaced by 0.25 mm.

(2) If the center is bright, we have

$$2d = m\lambda \, ,$$

where m is an integer.

For the first dark ring, we have

$$2d\cos\theta = \left(m - \frac{1}{2}\right)\lambda \, .$$

Subtracting yields

$$2d(1 - \cos\theta) = \frac{\lambda}{2} \, .$$

For small θ, $\cos\theta \approx 1 - \frac{\theta^2}{2}$, hence

$$\theta \approx \sqrt{\lambda/2d} \quad \text{rad} \, .$$

For $d = 0.25$ mm, we obtain $\theta = 0.032$ rad $= 1.8°$.

2011

Find the thickness of a soap film that gives constructive second order interference of reflected red light ($\lambda = 7000\overset{\circ}{A}$). The index of refraction of the film is 1.33. Assume a parallel beam of incident light directed at 30° to the normal.

(Chicago)

Solution:

Within a limited region the film can be considered as a parallel-sided slab of $n = 1.33$. The optical path difference of the beams reflected at the upper and the lower surfaces is

$$\Delta = 2nd\cos\theta + \frac{\lambda}{2} \, ,$$

For constructive second-order interference, we have

$$\Delta = 2\lambda \, ,$$

or

$$d = \frac{3\lambda}{4n\cos\theta},$$

where θ is the angle of refraction inside the film. From Snell's law, we have

$$\sin\theta_0 = n\sin\theta,$$

i.e.,

$$\sin\theta = \frac{1}{n}\sin\theta_0 = \frac{1}{2n},$$

or

$$\cos\theta = \sqrt{1 - \left(\frac{1}{2n}\right)^2},$$

which yields

$$d = 4260 \overset{\circ}{\text{A}}.$$

2012

A vertical soap film is viewed horizontally by reflected sodium light ($\lambda = 589 \times 10^{-9}$ m). The top of the film is so thin that it looks black in all colors. There are five bright fringes, the center of the fifth being at the bottom. How thick is the soap film at the bottom? The index of refraction of water is 1.33.

(*Wisconsin*)

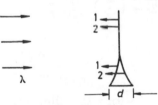

Fig. 2.14

Solution:

The thickness at the top of the film is much less than $\bar{\lambda}/(4n)$, where $\bar{\lambda}$ is the average wavelength of the visible spectrum of about 550 nm, and can be neglected. The phase difference between the rays (marked 1 and 2 in Fig. 2.14) reflected at the left and right surfaces then approaches π (additional phase shift on reflection), and the film appears black by reflected light.

The phase difference between the two beams at the bottom is

$$\delta = \pi + \frac{2nd}{\lambda} \cdot 2\pi,$$

where d is the thickness. For the centre of the fifth fringe, we have

$$\delta = 10\pi.$$

Using the given values of λ and n, we get

$$d = 1.0 \ \mu\mathrm{m}.$$

2013

White light is incident normally on a thin film which has $n = 1.5$ and a thickness of $5000\,\overset{\circ}{\mathrm{A}}$. For what wavelengths in the visible spectrum $(4000 - 7000\,\overset{\circ}{\mathrm{A}})$ will the intensity of the reflected light be a maximum?

(*Wisconsin*)

Solution:

Constructive interference occurs for

$$\mathrm{OPD} = 2nt = k\lambda + \frac{\lambda}{2},$$

where t is the thickness of the film, k an integer. Hence for maximum reflection the wavelengths are

$$\lambda = \frac{4nt}{2k+1} = \begin{cases} 6000\,\overset{\circ}{\mathrm{A}} & \text{for } k = 2 \\ 4285\,\overset{\circ}{\mathrm{A}} & \text{for } k = 3. \end{cases}$$

2014

A lens is to be coated with a thin film with an index of refraction of 1.2 in order to reduce the reflection from its surface at $\lambda = 5000\,\overset{\circ}{\mathrm{A}}$. The glass of the lens has an index of refraction of 1.4, as shown in Fig. 2.15.

(a) What is the minimum thickness of the coating that will minimize the intensity of the reflected light?

(b) In the above case the intensity of the reflected light is small but not zero. Explain. What needs to be changed, and by how much, to make the intensity of the reflected light zero?

(*Wisconsin*)

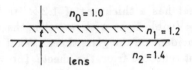

Fig. 2.15

Solution:

(a) If the light beams reflected from the two surfaces of the coating film are in antiphase,

$$2n_1 t = \lambda/2,$$

or $t = \lambda/(4n_1) = 0.10\,\mu\mathrm{m}$, the intensity of the reflected light will be minimized.

(b) When light is incident normally from medium A of refractive index n_A into medium B of n_B, the coefficient of reflection R is given by

$$R = [(n_A - n_B)/(n_A + n_B)]^2.$$

The coefficients of reflection at the upper and lower surfaces of the coating are respectively

$$R_1 = \left(\frac{n_1 - n_0}{n_1 + n_0}\right)^2 = \left(\frac{1}{1.1}\right)^2$$

and

$$R_2 = \left(\frac{n_1 - n_2}{n_1 + n_2}\right)^2 = \left(\frac{1}{1.3}\right)^2.$$

Because $R_1 \neq R_2$, the intensities of the light beams reflected from the two surfaces of the film are not identical and the intensity of the resultant light is not zero even though destructive interference occurs. If the index of refraction of the coating is n_1' so that $R_1 = R_2$, i.e.,

$$\left(\frac{n_1' - n_0}{n_1' + n_0}\right)^2 = \left(\frac{n_1' - n_2}{n_1' + n_2}\right)^2,$$

the intensity of the resultant reflected light would be zero. This requires that

$$n' = \sqrt{n_0 n_2} = 1.18.$$

2015

A thin glass sheet has a thickness of 1.2×10^{-6} m and an index of refraction $n = 1.50$. Visible light with wavelengths between 400 nm and 700 nm is normally incident on the glass sheet. What wavelengths are most intensified in the light reflected from the sheet? (nm$= 10^{-9}$ m)

(Wisconsin)

Solution:

Constructive interference occurs for

$$\text{OPD} = 2nd + \frac{\lambda}{2} = k\lambda \,,$$

where d is the thickness of the film, k an integer. Hence for maximum reflection,

$$\lambda = \frac{2nd}{k - \frac{1}{2}} \,.$$

In the wavelength range 400 to 700 nm, the most intensive reflected wavelengths are $\lambda_1 = 424$ nm (for $k = 9$), $\lambda_2 = 480$ nm (for $k = 8$), $\lambda_3 = 554$ nm (for $k = 7$), $\lambda_4 = 655$ nm (for $k = 6$).

2016

A TiO_2 film of index 2.5 is placed on glass of index 1.5 to increase the reflection in the visible. Choosing a suitable value for wavelength, how thick a layer in microns would you want, and what reflectivity would this give you?

(Wisconsin)

Solution:

Let n_0, n_1 and n_2, and d represent repectively the indices of refraction of air, TiO_2 film and glass, and the thickness of the film. Destructive interference between the reflected light occurs for

$$\text{OPD} = 2nd - \frac{\lambda}{2} = \left(k - \frac{1}{2} \right) \lambda \,,$$

where k is an integer, and the additional half wavelength is caused by the phase shift at the air-TiO_2 interface.

For $k = 0$ and $\lambda = 5500\overset{\circ}{A}$, we have $d = \lambda/(4n_1) = 0.055\,\mu$m. Hence the reflectivity is

$$R_{\lambda_0} = \left(\frac{n_0 - \frac{n_1^2}{n_2}}{n_0 + \frac{n_1^2}{n_2}}\right)^2 = 0.376\,.$$

2017

A soap film $(n = 4/3)$ of thickness d is illuminated at normal incidence by light of wavelength 500 nm. Calculate the approximate intensities of interference maxima and minima relative to the incident intensity as d is varied, when viewed in reflected light.

(*Wisconsin*)

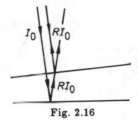

Fig. 2.16

Solution:

The reflectivity at each surface of the soap film is given by

$$R = \left[\frac{(n - n_0)}{(n + n_0)}\right]^2 = \left[\frac{\left(\frac{4}{3} - 1\right)}{\left(\frac{4}{3} + 1\right)}\right]^2 \approx 0.02\,.$$

For a film of low reflectivity R, the intensity of the reflected light at either surface can be approximately represented by RI_0, where I_0 is the intensity of the incident light. The intensity of the interference pattern is therefore

$$I = 2RI_0\,(1 + \cos\delta)\,,$$

where δ is the phase difference between the beams of reflected light at the two surfaces of the soap film given by

$$\delta = \left(\frac{2\pi d}{\lambda}\right)\cos\theta\,,$$

θ being the angle of incidence.

Hence

$$I \max / I_0 = 4R = 0.08 ,$$
$$I_{min}/I_0 = 0 .$$

2018

Explain how the colored rings in the demonstration experiment shown in Fig. 2.17 are formed, and show that the estimated size of the rings is consistent with your explanation. Why are the rings so bright?

Hint: Mirror problems involve "folded optics", unfold this problem.

(*Princeton*)

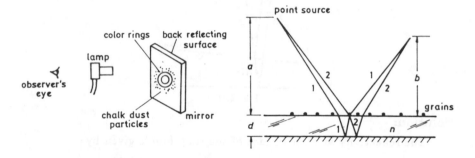

Fig. 2.17   Fig. 2.18

Solution:

Consider two coherent rays leaving a point source as shown in Fig. 2.18. Ray 1 is reflected by the mirror and then scattered by a grain towards a certain point in space. Ray 2 is first scattered by the grain towards the mirror and then reflected back towards the abovementioned point.

The resulting OPD determines the interference at the point. At normal incidence, the pattern is a series of concentric rings of radii r:

$$r = \left(\frac{nm\lambda a^2 b^2}{d(a^2 - b^2)} \right)^{\frac{1}{2}} ,$$

where n is the index of refraction of the glass, d the thickness of the glass slab, λ the wavelength of light, a the distance between the slab and the point source, b the distance between the slab and the interference point, m an integer.

For white light source, color rings appear as different frequencies give rise to different rings.

The rings are bright because of constructive interference.

2019

A 5 mm thick glass window 2 cm in diameter is claimed to have each surface flat to within 1/4 wave of mercury green light ($\lambda = 546$ nm), and the pair of surfaces parallel to within 5 seconds of arc (1 arc sec $= 4.85 \times 10^{-6}$ radians). How would you measure these properties to verify the manufacturer's specifications? Your may assume the refractive index of the glass is $n = 1.500$.

(*Wisconsin*)

Solution:

To test the flatness of a surface, we place a plano-convex lens of long focal length on it with the convex side down. We then illuminate it from above with the mercury green light as shown in Fig. 2.19 and observe the Newton's rings formed. The spacing between adjacent rings corresponds to a change of thickness of the air gap of $\frac{\lambda}{2}$. For flatness to within $\frac{\lambda}{4}$, the distortion of a ring should be less than one-half of the spacing.

To test the parallelness of the surfaces of the glass window we consider it as a wedge of glass of refractive index $n = 1.500$. Illuminating it from the above, we shall observe interference fringes such that in going from one fringe to the next, the thickness d changes by $\frac{\lambda}{2n}$. For a wedge angle θ, the spacing is

$$S = \frac{\Delta d}{\theta} = \frac{\lambda}{2n\theta}.$$

For $\theta \lesssim 5 \times 4.85 \times 10^{-6}$ rad, we require that

$$S \gtrsim \frac{546 \times 10^{-7}}{2 \times 1.5 \times 24.25 \times 10^{-6}} = 0.75 \text{ cm}.$$

Both of the above conditions should be met to justify the manufacturer's claims.

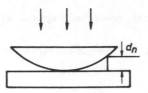

Fig. 2.19

2020

The radius of curvature of the convex surface of a plano-convex lens is 30 cm. The lens is placed with its convex side down on a plane glass plate, and illuminated from above with red light of wavelength 650 nm (Fig. 2.20).

(a) Find the diameter of the third bright ring in the interference pattern.

(b) Show that for large R this diameter is approximately proportional to $R^{1/2}$.

(SUNY, Buffalo)

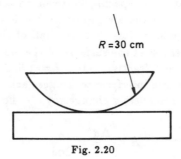

$R = 30$ cm

Fig. 2.20

Solution:

(a) The radii of the bright Newton's rings are given by the following formula,

$$r_j = \sqrt{(2j+1)\frac{\lambda}{2}R} \qquad (j = 0, 1, 2, \ldots).$$

For the third bright ring, $j = 2$, and the radius is

$$r = \sqrt{(4+1) \times \frac{650 \times 10^{-7}}{2} \times 30} = 0.7 \text{ mm},$$

i.e., the diameter is

$$d = 2r = 1.4 \text{ mm}.$$

(b) From the formula we see that $d = 2r \propto \sqrt{R}$. It should be noted that in deriving the formula we have already used an approximation based on large R.

2021

A double-slit interference is produced by plane parallel light of wavelength λ passing through two slits of unequal widths, say $w_1 = 20\lambda$ and $w_2 = 40\lambda$, their centers being 1000λ apart. If the observation is made on a screen very far away from the slits, i.e., at a distance $L \gg 1000\lambda$, determine the following features:

(a) Separation δx between adjacent maxima.

(b) Widths Δx_1 and Δx_2 of the central maxima of the diffraction patterns of the two slits individually (i.e., distances between the first zeros).

(c) Hence, the number of fringes produced by the overlap of these central maxima.

(d) The intensity ratio between intensity maximum and minimum in the center of the pattern.

(e) An analytic expression for the intensity on the screen as a function of x when $x = 0$ is at the exact center of the pattern.

Show your reasoning.

(UC, Berkeley)

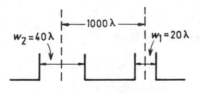

Fig. 2.21

Solution:

(a) As $w_2 = 2w_1, L \gg w_1, w_2$, we may take the amplitudes of the electric field vectors produced by the slits S_1 and S_2 individually on the screen to be E and $2E$ respectively. Since $I \propto E^2$, the corresponding intensities are $I_1 = I_0$, say, $I_2 = 4I_0$.

For double-slit interference the resultant amplitude is given by the vector sum of E_1 and E_2 making an angle δ equal to the phase difference. Hence

$$I = I_1 + I_2 + 2\sqrt{I_1 I_2}\cos\delta$$
$$= I_0 + 4I_0 + 4I_0 \cos(2\pi\,d\sin\theta/\lambda) = 5I_0 + 4I_0 \cos(2000\pi\sin\theta),$$

where d is the separation of the centres of the slits, being equal to 1000λ, $\sin\theta \approx \frac{x}{L}$, x being the distance on the screen from the centre of the interference pattern. As $L \gg x$, $\sin\theta \simeq \theta$ and adjacent maxima occur when $2000\pi\delta\theta = 2\pi$, or $\delta\theta = \frac{1}{1000}$, giving

$$\delta x = L\delta\theta = \frac{L}{1000}.$$

(b) The diffraction intensity distribution due to a slit of width w is given by $\sim \left(\frac{\sin\beta}{\beta}\right)^2$, where $\beta = \frac{\pi w}{\lambda}\sin\theta$. Thus

$$I_1 = I_0 \left(\sin\beta_1/\beta_1\right)^2, \quad \text{where } \beta_1 = \pi w_1 \sin\theta/\lambda = 20\pi\sin\theta,$$
$$I_2 = 4I_0 \left(\sin\beta_2/\beta_2\right)^2, \quad \text{where } \beta_2 = \pi w_2 \sin\theta/\lambda = 40\pi\sin\theta.$$

The first zeros occur at angular distances from the centre of θ_1 and θ_2 given by

$$20\pi\sin\theta_1 \simeq 20\pi\theta_1 = \pi, \text{ or } \theta_1 = \frac{1}{20}, \text{ and } \theta_2 = \frac{1}{40}.$$

Hence the angular widths of the central maxima are

$$\Delta\theta_1 = 2\theta_1 = \frac{1}{10},$$
$$\Delta\theta_2 = \frac{1}{20}.$$

The corresponding widths are then

$$\Delta x_1 = L\theta_1 = L/10,$$
$$\Delta x_2 = L\theta_2 = L/20.$$

(c) The width of the overlap of these central maxima is $L/20$, and the number of fringes is $(L/20)/(L/1000)=50$.

(d) The intensity maximum and minimum in the center of the pattern are given by $\cos\delta = 1$ and $\cos\delta = -1$ respectively. Hence we have the ratio

$$\frac{I_{\max}}{I_{\min}} = \frac{9I_0}{I_0} = 9.$$

(e) Taking diffraction effect into account gives the fields produced by S_1 and S_2 respectively as

$$E_1 = E_0(\sin \beta_1/\beta_1),$$
$$E_2 = 2E_0(\sin \beta_2/\beta_2),$$

where E_0 is a constant.

With $\beta_2 = 2\beta_1$, we have

$$E = |\mathbf{E}_1 + \mathbf{E}_2| = E_0 \frac{\sin \beta_1}{\beta_1} \left(5 + 4\cos \delta\right)^{\frac{1}{2}}$$

The intensity is therefore

$$I(x) = I_0 \left(\frac{\sin \frac{20\pi x}{L}}{\frac{20\pi x}{L}}\right)^2 \left(5 + 4\cos \frac{2000\pi x}{L}\right).$$

2022

The diagram (Fig. 2.22) shows a double-slit experiment in which coherent monochromatic light of wavelength λ from a distant source is incident upon the two slits, each of width $w(w \gg \lambda)$, and the interference pattern is viewed on a distant screen. A thin piece of glass of thickness δ, index of refraction n, is placed between one of the slits and the screen perpendicular to the light path, and the intensity at the central point P is required as a function of thickness δ. Assume that the glass does not absorb or reflect any light. If the intensity for $\delta = 0$ is given by I_0:

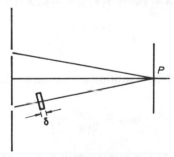

Fig. 2.22

(a) What is the intensity at point P as a function of thickness δ?

(b) For what values of δ is the intensity at P a minimum?

(c) Suppose the width w of one of the slits is now increased to $2w$, the other width remaining unchanged. What is the intensity at point P as a function of δ?

(*Columbia*)

Solution:

(a) For the double slit system, the intensity at a point on the screen is given by

$$I \sim 4a^2 \left(\frac{\sin \beta}{\beta} \right)^2 \cos^2 \left(\frac{\varphi}{2} \right),$$

where a is the amplitude due to each individual slit, $4a^2(\sin \beta/\beta)^2$ is the single-slit diffraction term, $\cos^2 (\varphi/2)$ is the interference term, with

$$\varphi = \frac{2\pi}{\lambda}(n - 1)\,\delta$$

being the phase difference between the waves originated from the two slits, n the index of refraction of the glass and δ the thickness of the glass plate. The intensity at the central point P is given by $\beta \to 0$, for which $\frac{\sin \beta}{\beta} \to 1$, as

$$I \sim 4a^2 \cos^2 \left(\frac{\pi}{\lambda}(n - 1)\delta \right),$$

or

$$I = I_0 \cos^2 \left(\frac{\pi}{\lambda}(n - 1)\delta \right),$$

where I_0 is the intensity at P for $\delta = 0$.

(b) For $\pi\delta(n - 1)/\lambda = (2k + 1)\pi/2, k = 0, \pm 1, \pm 2, \ldots$, or

$$\delta = \frac{(2k + 1)\lambda}{2(n - 1)},$$

the intensity at P would be a minimum. In particular for $k = 0$, we get

$$\delta_{\min} = \frac{\lambda}{2(n - 1)}.$$

Note that the intensity at the minima will not be exactly zero as the slits have finite width.

(c) As the width of one of the slits is increased to $2w$, the amplitude due to the slit will become $2a$. The intensity of the resultant wave at P is now given by

$$
\begin{aligned}
I &\sim a_1^2 + a_2^2 + 2a_1 a_2 \cos\varphi \\
&= a^2 + (2a)^2 + 2a \cdot (2a)\cos\varphi \\
&= a^2(5 + 4\cos\varphi) = a^2[5 + 4\cos(2\pi(n-1)\delta/\lambda)] .
\end{aligned}
$$

For $\delta = 0$, the intensity at P is the maximum $I_0' = 9a^2$. Comparing with (a), we have

$$
\frac{I_0'}{I_0} = \frac{9a^2}{4a^2} = 2.25 .
$$

2023

A Young double-slit interferometer receives light from a stellar object which is focused at the plane shown in Fig. 2.23.

(a) Determine the interference pattern as a function of x.

(b) If the interferometer is pointed at a stellar object subtending, at the earth, an angle greater than $\theta_{\min}$, the pattern disappears. Explain why and find $\theta_{\min}$.

(Columbia)

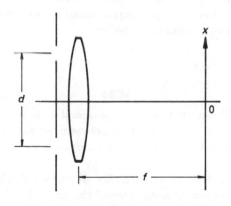

Fig. 2.23

Solution:

(a) Let the width of each slit be a and the angular size of the star be θ. To good approximation the intensity distribution of the interference pattern is given by the result of Question 2007:

$$I(x, \theta) \sim \left(\frac{\sin \beta}{\beta}\right)^2 \left(1 + \frac{\sin \alpha}{\alpha} \cos \gamma\right)$$

where

$$\beta = \frac{\pi a x}{\lambda f}, \alpha = \frac{\pi d \theta}{\lambda}, \gamma = \frac{2\pi d x}{\lambda f}.$$

(b) The visibility of the pattern is given by

$$V = \frac{I_{\max} - I_{\min}}{I_{\max} + I_{\min}},$$

where $I_{\max}$ and $I_{\min}$ are the intensities at the neighboring maximum and minimum respectively. Thus

$$V = \frac{\sin \alpha}{\alpha}.$$

As θ increases from 0 to $\theta_{\min}$, V decreases from 1 to 0, where $\theta_{\min}$ is given by

$$\frac{\pi d \theta_{\min}}{\lambda} = \pi, \text{ or } \theta_{\min} = \frac{\lambda}{d}.$$

The pattern also appears when $\theta \geq \theta_{\min}$. Actually, beyond $\theta_{\min}$ the pattern may reappear and then disappear again many times but the fluctuation of V around 0 becomes smaller and smaller.

2024

A two-slit diffraction system is illuminated normally with coherent light of wavelength λ. The separation between the slits is a. The intensity of light detected at a distant screen (far compared with a), when one of the slits is covered, is I_0.

(a) For the situation where both slits are open, calculate and sketch the response of the detector as a function of the angle θ, where θ is measured away from the normal to the system.

(b) Now assume that the slit separation "jitters" such that the spacing a between the slits changes on a timescale large compared to the period of the light but short compared with the response time of the intensity detector. Assume that the spacing a has a Gaussian probability distribution of width Δ and average value $\bar{a}$. Assume that $\bar{a} \gg \Delta \gg \lambda$. Without performing a detailed calculation, sketch the intensity pattern measured by the detector for this situation.

(*MIT*)

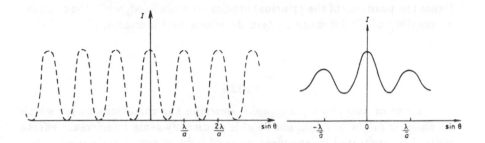

Fig. 2.24   Fig. 2.25

Solution:

(a) Assume the widths of the slits are much smaller than λ, then one need not consider the diffraction produced by each slit. Only the interference between beams from different slits needs to be considered.

The amplitudes of the waves at a point on the screen produced individually by the slits are given by

$$E_1 = Ae^{ikr_1}$$

and

$$E_2 = Ae^{ikr_2} = Ae^{ikr_1} \cdot e^{ika \sin \theta},$$

where $k = \frac{2\pi}{\lambda}$, r_1 and r_2 are the distances between the point and the two slits. The resultant intensity is

$$I \sim EE^* = 2A^2[1 + \cos(ka \sin \theta)]$$
$$= 4A^2 \cos^2 \left(\frac{\pi a \sin \theta}{\lambda} \right)$$

and is plotted in Fig. 2.24.

(b) Since the slit separation changes on a time-scale small compared with the response time of the intensity detector, the intensity pattern will reflect the probability distribution of the slit separation which is Gaussian:

$$p(a) = \frac{1}{\sqrt{2\pi}\Delta} e^{-\frac{(a-\bar{a})^2}{2\Delta^2}}.$$

Thus the intensity detected will distribute as

$$I \sim p(a) \cos^2 \left(\frac{\pi a \sin \theta}{\lambda} \right).$$

Hence the positions of the principal maxima remain fixed, while the contrast or visibility of the intensity pattern decreases as θ increases, (Fig. 2.25).

2025

In one or two short paragraphs, describe the conditions under which an observer might or might not be able to directly sense interference effects created by separated independent pairs of light or sound generators.

(*UC, Berkeley*)

Solution:

An observer might be able to directly sense interference effects if two independent light or sound generators are in phase during a period of time which is longer than the response time of human sense organs (eyes or ears) and the powers of the two generators are comparable to each other.

Common light emitters and sound generators do not satisfy the former condition. So one seldom observes directly interference effects. For laser light of high coherence, one can directly see interference effects.

2026

(a) Distinguish between Fraunhofer and Fresnel diffractions in terms of the experimental arrangement used.

(b) Show schematically an experimental arrangement which will allow Fraunhofer diffraction to be observed.

(c) Draw the pattern observed on a screen of the Fraunhofer diffraction from a single slit (width a) and from a double slit (widths a, separation d). Point out the distinguishing features of each pattern.

(d) Calculate the interference pattern that would be obtained if three equally spaced slits were used instead of two in Young's experiment (screen far from slits).

(*SUNY, Buffalo*)

Solution:

(a) Fraunhofer diffraction is the diffraction observed when the source of light and the screen on which the pattern is observed (Fig. 2.26) are effectively at infinite distances from the aperture causing the diffraction so that both the incident and diffracted beams can be regarded as plane waves. Experimentally the light from a source is rendered parallel with a lens and is focused on the screen with another lens placed behind the aperture. Fresnel diffraction is that observed when either the source or the screen, or both, are at finite distances from the aperture. Here the wave fronts are divergent instead of plane. No lenses are necessary in the observation of Fresnel diffraction.

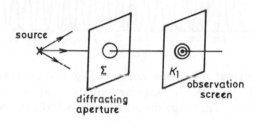

Fig. 2.20

(b) A light source is located on the front focal plane of a convex lens L_1, the light emerging from which is in the form of plane waves. The waves fall on a diffracting aperture, behind which a convex lens L_2 is located. On the back focal plane of L_2 is an observation screen, on which the Fraunhofer diffraction pattern appears. See Fig. 2.27.

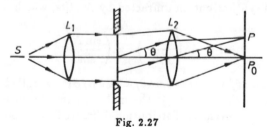

Fig. 2.27

(c) The diffraction intensity distribution from a single-slit of width a is given by

$$I \sim A_0^2 \left(\frac{\sin \beta}{\beta} \right)^2$$

where $\beta = \pi a \sin \theta / \lambda$. The intensity distribution is shown in Fig. 2.28.

The features are as follows.

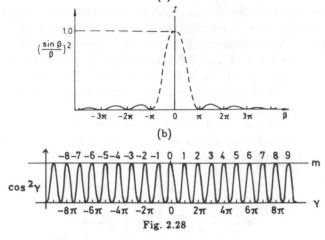

Fig. 2.28

1. The principal maximum occurs when θ approaches zero, so that $(\sin\beta/\beta)^2$ becomes unity. Its angular half-width at the aperture is given by $\sin\beta = \pi$ or $\theta \simeq \lambda/a$. Thus the smaller the width, the wider will be the diffraction pattern.

2. The diffraction intensity is zero at angles (except 0) for which $\sin\beta = 0$, or $\sin\theta = k\lambda/a \, (k = 1, 2, \ldots)$. Between two neighboring minima there is a weak maximum, the intensity of which is much less than that of the principal maximum, attenuating as k increases.

3. The angular width of the central principal intensity maximum is twice that of the other maxima.

The intensity distribution diffracted by N slits, width a and spacing d is given by

$$I = A_0^2 \left(\frac{\sin\beta}{\beta}\right)^2 \times \left(\frac{\sin N\gamma}{\sin\gamma}\right)^2$$

where $\beta = \pi a \sin\theta/\lambda, \gamma = \pi d \sin\theta/\lambda$. $(\sin\beta/\beta)^2$ is called the single-slit diffraction factor, $(\sin N\gamma/\sin\gamma)^2$ the multi-slit interference factor.

For $N = 2$, the intensity distribution diffracted by a double-slit is given by

$$I = 4A_0^2 (\sin\beta/\beta)^2 \cos^2\gamma.$$

The intensity pattern diffracted by a double-slit of spacing $d \, (d = 3a$, for instance) is shown in Fig. 2.29, which is the intensity pattern of double-slit interference (Fig. 2.28(b)) modulated by the envelope of the intensity pattern of single-slit diffraction. (Fig. 2.28(a)). The maxima occur if $\gamma = m\pi$ or $\sin\theta = m\lambda/d$, where $m = 0, \pm 1, \pm 2, \ldots$ is the order of interference.

The minima of the interference pattern are given by $\sin\theta = (m + \frac{1}{2})\lambda/d$. The diffraction minima are given by $\sin\theta = k\lambda/a$ $(k = \pm1, \pm2, \dots)$ at which the diffraction factor becomes zero. An effect known as missing orders can occur if $d/a = m/k$ is a rational fraction. Then the m-th interference maximum and the k-th diffraction minimum (zero) correspond to the same θ value and that maximum will not appear, or at least reduced to a very low intensity. For $d = 3a$, the third, sixth, ninth, $\dots$ orders will not appear, as seen in Fig. 2.29.

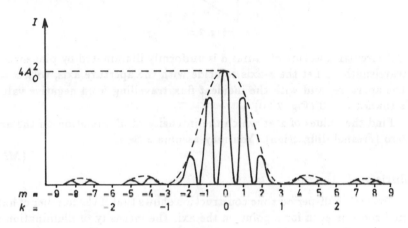

Fig. 2.29

(d) The 3-slit diffraction intensity is given by

$$I = A_0^2 \frac{\sin^2\left(\frac{\pi a \sin\theta}{\lambda}\right)}{\left(\frac{\pi a \sin\theta}{\lambda}\right)^2} \frac{\sin^2\left(\frac{3\pi d \sin\theta}{\lambda}\right)}{\sin^2\left(\frac{\pi d \sin\theta}{\lambda}\right)} .$$

2027

Show that Fresnel half-period zones for a circular aperture all have equal areas.

(Wisconsin)

Solution:

The proof can be found in most optics textbooks and is therefore omitted here.

2028

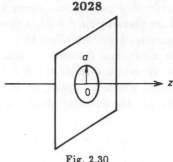

Fig. 2.30

A circular aperture of radius a is uniformly illuminated by plane waves of wavelength λ. Let the z-axis coincide with the aperture axis, with $z = 0$ at the aperture, and with the incident flux travelling from negative values of z toward $z = 0$ (Fig. 2.30).

Find the values of z at which the intensity of illumination on the axis is zero (Fresnel diffraction). You may assume $z \gg a$.

(*MIT*)

Solution:

Fresnel's half-period zone construction shows that if the number of half-period zones is even for a point on the axis the intensity of illumination at this point will be zero. As

$$N = \frac{a^2}{\lambda z},$$

or

$$z = \frac{a^2}{\lambda N},$$

the intensity at the point z will be zero for $N = 2, 4, 6, \ldots$.

2029

Monochromatic light of wavelength λ is normally incident on a screen with a circular hole of radius R. Directly behind the hole on the axis the intensity of light vanishes at some points; how far from the screen is the most distant point at which the light vanishes? Assume $\lambda/R \ll 1$.

(*Wisconsin*)

Solution:

The most distant point at which the light vanishes is at $R^2/(2\lambda)$ from the screen.

2030

Light of wavelength λ and intensity I_0 is incident perpendicularly onto
(a) an opaque circular disk of radius R,
(b) an opaque screen with a circular hole of radius R.

In each case, find the intensity of the light at a distance L behind the obstacle, on the path passing through the center of the circle. Take $L \gg R$.

(*Wisconsin*)

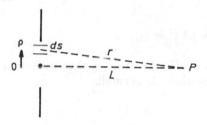

Fig. 2.31

Solution:

Let us first consider the diffraction by the hole (see Fig. 2.31). The field at P on the axis, distance L from the hole, is represented by

$$A = A_0 \int \int \frac{K e^{ikr} ds}{r},$$

where K is the obliquity factor, which can be taken as $1/i\lambda$ for $L \gg R$, $k = 2\pi/\lambda$ is the wave number, ds is the annular elemental area of radius ρ, $ds = 2\pi\rho d\rho$, and

$$r = \sqrt{\rho^2 + L^2}$$

is the distance between P and ds. As $L \gg R \geq \rho$, r can be taken as L in the denominator and $L(1 + \rho^2/2L^2)$ in the exponent, which is the Fresnel approximation. We then have

$$A = \frac{A_0}{i\lambda} \frac{e^{ikL}}{L} \int_0^R 2\pi\rho e^{ik\rho^2/2L} d\rho$$

$$= A_0 \left[e^{ikL} - e^{ik(L + \frac{R^2}{2L})} \right],$$

where we have omitted some constant factor. The intensity is therefore

$$I \sim |A|^2 = 4I_0 \sin^2(kR^2/4L),$$

where $I_0 = |A_0|^2$.

For diffraction by the opaque circular disk, the field A' at P is given by Babinet's principle

$$A + A' = A_\infty = A_0 e^{ikL},$$

where A is the field at P due to diffraction by the circular hole above, A_∞ is the field due to diffraction by a hole of infinite size. Thus

$$A' = A_0 e^{ikL} - A = A_0 \exp\left[ik\left(L + \frac{R^2}{2L}\right)\right],$$

giving

$$I' \sim |A'|^2 = I_0,$$

i.e., the intensity is nearly the same as that without any diffracting screen provided the opaque obstacle is small.

2031

Monochromatic parallel light impinges normally on an opaque screen with a circular hole of radius r and is detected on the axis a distance L from the hole. The intensity is observed to oscillate as r is increased from 0 to ∞.

(a) Find the radius r_a of the hole for the first maximum.

(b) Find the radius r_b of the hole for the first minimum.

(c) Find the ratio of the intensity for $r = r_a$ to the intensity for $r = r_\infty$.

(d) Suppose the screen is replaced by on opaque disk of radius r_a. What is the intensity?

(*Chicago*)

Solution:

For parallel incident light and a point on the axis distance L from the hole, the radii of the Fresnel zones are given by

$$r_k = \sqrt{k\lambda L}, \quad (k = 1, 2, 3, \dots).$$

(a) For the first maximum, $k = 1, r_a = r_1 = \sqrt{\lambda L}$.

(b) For the first minimum, $k = 2, r_b = r_2 = \sqrt{2\lambda L}$.

(c) As k approaches infinity, the total amplitude is

$$A_\infty = A_1 - A_2 + A_3 - A_4 + \ldots = \frac{A_1}{2} + \left(\frac{A_1}{2} - A_2 + \frac{A_3}{2}\right)$$

$$+ \left(\frac{A_3}{2} - A_4 + \frac{A_5}{2}\right) + \ldots \approx \frac{A_1}{2},$$

as

$$A_2 \approx \frac{1}{2}(A_1 + A_3), \ A_4 \approx \frac{1}{2}(A_3 + A_5), \ldots,$$

$$I_\infty \sim A_\infty^2 = A_1^2/4 \sim I_1/4,$$

where I_1 is the intensity produced by the first half-period zone at the point in question. In other words, the ratio of the intensity for $r = r_a$ to the intensity for $r = r_\infty$ is 4.

(d) From Babinet's principle, the sum of field amplitudes A_1 and A_1' produced by a pair of complementary screens equals A_∞, amplitude of the unscreened wave, i.e., $A_1 + A_1' = A_\infty$. A_1' is the field amplitude produced by an opaque disk of radius r_a. Therefore the intensity is

$$I' \sim (A_\infty - A_1)^2 \sim I_1/4 = I_\infty.$$

2032

An opaque sheet has a hole in it of 0.5 mm radius. If plane waves of light ($\lambda = 5000\ \overset{\circ}{A}$) fall on the sheet, find the maximum distance from this sheet at which a screen must be placed so that the light will be focused to a bright spot. What is the intensity of this spot relative to the intensity with the opaque sheet removed? Indicate how you arrived at each answer.

(*Wisconsin*)

Solution:

The maximum distance r of the screen from the hole for which the light will be focused to a bright spot is that for which the area of the hole corresponds to the first Fresnel zone only, and is given by

$$\rho_1^2 = \lambda r,$$

where ρ is the radius of the hole. Hence

$$r = \frac{\rho_1^2}{\lambda} = \frac{0.5^2}{5000 \times 10^{-7}} = 500 \text{ mm}.$$

Without the opaque screen, the number of the Fresnel zones will approach infinity. The total amplitude is then

$$A = A_1 - A_2 + A_3 - \ldots\ldots$$
$$= \frac{A_1}{2} + \left(\frac{A_1}{2} - A_2 + \frac{A_3}{2}\right) + \left(\frac{A_3}{2} - A_4 + \frac{A_5}{2}\right) + \ldots$$
$$\approx \frac{A_1}{2},$$

hence

$$I' \sim A_1^2/4, \text{ i.e., } I' = I_1/4.$$

Thus the intensity is now $1/4$ that before the opaque sheet is removed.

2033

A plane wave of microwaves having $\lambda = 1$ cm is incident on an opaque screen containing a circular aperture of adjustable radius. A detector having a very small sensitive area is located on the axis of the aperture 1 meter behind it. If the radius of the aperture is gradually increased from zero, at what value of the radius does the detector response reach its first maximum? Its second minimum after the first maximum? At the latter radius find the positions of the maxima and minima along the axis.

(*Wisconsin*)

Solution:

For incident plane waves the radii of the Fresnel zones are given by $\rho_m = \sqrt{m r_0 \lambda}$, $m = 1, 2, 3, \ldots$ where r_0 is the distance of the detector from the aperture. For the first maximum, $m = 1$, $\rho_1 = 0.1$ m.

For $m = 4$, when the hole area corresponds to 4 Fresnel zones, the detector response reaches its second minimum. Thus

$$\rho_4 = \sqrt{4 r_0 \lambda} = 0.2 \text{ m}.$$

At the radius 0.2 m, the positions of the maxima and minima can be found from the formula

$$\rho_4 = \sqrt{k r_k \lambda}, \quad (k = 1, 2, 3, \ldots\ldots),$$

or $r_k = \rho_4^2/(k\lambda)$.

For k an odd integer, $r_1 = 4$ m, $r_3 = 1.33$ m, $r_5 = 0.8$ m... the maxima occur; for k an even integer, $r_2 = 2$ m, $r_4 = 1$ m, $r_6 = 0.67$ m,... the minima occur.

2034

A Fresnel zone plate is made by dividing a photographic image into 5 separate zones.

The first zone consists of an opaque circular disc of radius r_1. The second is a concentric transparent ring from r_1 to r_2 followed by an opaque ring from r_2 to r_3, a second transparent ring from r_3 to r_4 and a final zone from r_4 to infinity that is black. The radii r_1 to r_4 are in the ratio $r_1 : r_2 : r_3 : r_4 = 1 : \sqrt{2} : \sqrt{3} : \sqrt{4}$.

The zone plate is placed in the x-y plane and illuminated by plane monochromatic light waves of wavelength 5,000 Å. The most intense spot of light behind the plate is seen on the axis of the zone plate 1 meter behind it.

(a) What is the radius r_1?

(b) What is the intensity at that spot in terms of the intensity of the incident wave?

(c) Where can you expect to find the intensity maxima of the axis?

(*UC, Berkeley*)

Solution:

(a) The focal lengths of the zones are given by

$$f = \frac{r_j^2}{j\lambda}.$$

For $j = 1, f = 1$ m, $\lambda = 5000$ Å, we have $r_1 = 0.707$ mm.

(b) The resultant amplitude produced by the second and the fourth transparent zones is

$$A = A_2 + A_4 \approx 2A_1$$

where A_1 is the field amplitude produced by the first zone alone if it were transparent. As $A_1 = 2A_\infty$, A_∞ being the field at the point with the plate removed, we obtain

$$I \sim A^2 \approx 16A_\infty^2 \sim 16I_0.$$

(c) The Fresnel zone plate has a series of focal lengths which are given by

$$f_m = \frac{1}{m}\left(\frac{r_1^2}{\lambda}\right)$$

where m is an odd number. So we can expect to find intensity maxima on the axis at points distance $1/3$ m, $1/5$ m, $1/7$ m,... from the zone plate.

2035

Light from a monochromatic point source of wavelength λ is focused to a point image by a Fresnel half-period zone plate having 100 open odd half-period zones $(1, 3, 5, \ldots 199)$ with all even zones opaque. Compare the image dot intensity with that at the same point for the zone plate removed, and for a lens of the same focal length and diameter corresponding to 200 half-period zones of the zone plate. Assume the diameter of the opening is small compared to the distance from the source and the image.

(*Columbia*)

Solution:

Let the amplitudes of the light from the half-period zones be $A_1, A_3, \ldots$ A_{199}. As all of them are approximately the same as A_1, the resultant amplitude for the zone plate is given by

$$A = A_1 + A_3 + \ldots + A_{199} \approx 100A_1.$$

The intensity is then

$$I \sim A^2 \sim 10^4 \times A_1^2.$$

When we substitute a lens of the same focal length for the zone plate, all the waves of amplitudes $A_1, A_2, \ldots A_{199}, A_{200}$ which are in phase are focused. So the resultant amplitude A' is

$$A' = \sum_{i=1}^{200} A_i \approx 200\, A_1,$$

giving

$$I' \sim 4 \times 10^4 \times A_1^2.$$

Hence

$$I'/I = 4.$$

The intensity with the lens is 4 times that with the zone plate.

2036

(a) A coherent plane wave of wavelength λ is incident upon a screen with an infinitely long slit of width d. The beam's propagation direction is perpendicular to the slit and makes an angle θ with the normal to the screen (Fig. 2.32). Find the intensity distribution beyond the slit in the focal plane of a thin lens of focal length f. [Assume the lens, diameter infinite.]

(b) If, instead of a screen with a slit, there is a slab of width d (Fig. 2.33), find the intensity distribution in the focal plane.

(Columbia)

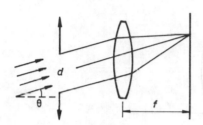 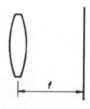

Fig. 2.32   Fig. 2.33

Solution:

(a) Using the coordinates shown in Fig. 2.34, the incident plane wave is represented by

$$e^{i\frac{2\pi}{\lambda}y' \sin \theta} .$$

The amplitude distribution in the focal plane of the lens is the amount of Fraunhofer diffraction by the slit:

$$A \propto \int_{-\frac{d}{2}}^{\frac{d}{2}} e^{i\frac{2\pi}{\lambda} \sin \theta y'} e^{-i 2\pi f_y y'} \, dy'$$

$$\propto \frac{\sin \left[\pi \left(\frac{\sin \theta}{\lambda} - f_y \right) d \right]}{\pi \left(\frac{\sin \theta}{\lambda} - f_y \right) d}$$

with

$$f_y = \frac{y}{\lambda f} .$$

Hence

$$I = a^2 \frac{\sin^2 \left[\pi \left(\frac{\sin \theta}{\lambda} - \frac{y}{\lambda f} \right) d \right]}{\left[\pi \left(\frac{\sin \theta}{\lambda} - \frac{y}{\lambda f} \right) d \right]^2} ,$$

where $a =$ constant.

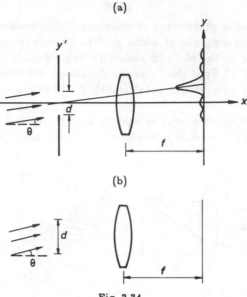

Fig. 2.34

The principal maximum is given by $\frac{\sin\theta}{\lambda} - \frac{y}{\lambda f} = 0$, or $y = f\sin\theta$. The minima are given by $\pi\left(\frac{\sin\theta}{\lambda} - \frac{y}{\lambda f}\right)d = k\pi$, or $y = f\sin\theta - \frac{k\lambda f}{d}$, where $k = \pm1, \pm2, \ldots$.

(b) If we substitute a complementary slab of the same width d for the screen with a slit, then according to Babinet's principle the intensity distribution in the focal plane remains unchanged except in the neighborhood of the central point $y = f\sin\theta$, as shown in Fig. 2.35.

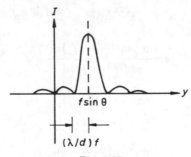

Fig. 2.35

2037

A plane wave with wavenumber k is incident on a slit of width a. The slit is covered by a transparent wedge whose thickness is proportional to the distance from the top of the slit $(t = \gamma x)$, see Fig. 2.36. The index of refraction of the transparent wedge is n. The intensity of light at an angle θ is given by

$$I \propto \sin^2(\beta a)/(\beta a)^2 .$$

Derive an expression for β in terms of k, n, γ and θ.

(*Wisconsin*)

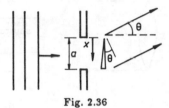

Fig. 2.36

Solution:

For a wedge with thickness $t = \gamma x$, the apex angle (in radians) is γ to a good approximation. The angle of deviation of light passing through the wedge is then

$$\delta = (n - 1)\gamma .$$

The intensity distribution due to single-slit diffraction is

$$I \propto \frac{\sin^2\left(\frac{1}{2}ak\sin\theta\right)}{\left(\frac{1}{2}ak\sin\theta\right)^2}$$

where $k = 2\pi/\lambda$ and θ is the angle made by the diffracted light with the axis. For the slit covered by the wedge, we should substitute $\theta - \delta$ for θ in above formula, i.e.,

$$I \propto \frac{\sin^2\left\{\frac{1}{2}ak\sin\left[\theta - (n-1)\gamma\right]\right\}}{\left\{\frac{1}{2}ak\sin[\theta - (n-1)\gamma]\right\}^2} .$$

Hence

$$\beta = \frac{1}{2}k\sin[\theta - (n-1)\gamma] .$$

Thus the distribution remains that for the single-slit diffraction, except that the center is displaced by an angle $(n-1)\gamma$.

2038

Suppose you wish to observe reflection of laser light from the moon. Explain how you could use a large-aperture telescope lens or mirror, together with a laser and a short-focus lens to get a beam with very small divergence. What would the characteristics of the telescope lens and short-focus lens have to be to get a beam of angular width 10^{-6} radian (first minimum on one side to first minimum on the other)? You may assume a rectangular aperture at the objective if you wish.

(*Wisconsin*)

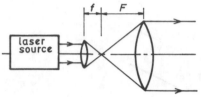

Fig. 2.37   Fig. 2.38

Solution:

The divergence of a light beam is caused mainly by aperture diffraction, the angular divergence being inversely proportional to the radius of the aperture. Thus the divergence may be reduced by expanding the cross section of the beam. Figures 2.37 and 2.38 show the arrangements for achieving this purpose, using respectively a large-aperture lens and a mirror. They may be considered as telescope used in reverse.

The laser beam from the source has radius r and hence angular divergence

$$\theta_1 = 1.22\frac{\lambda}{r}\,.$$

We use a short-focus lens and a large aperture lens (or mirror) of radii and focal length r, f and R, F respectively. The angular divergence of light from the lens system is

$$\theta_2 = \frac{\theta_1 f}{F} = \frac{1.22\lambda f}{rF}\,,$$

while that from the mirror is

$$\theta_3 = \frac{1.22\lambda}{R}\,.$$

For $\theta_2, \theta_3 \leq 10^{-6}$, if we assume $\lambda = 5000\,\overset{\circ}{\mathrm{A}}$ and $r = 6$ mm, we require $\frac{F}{f} \geq 100$ or $R \geq 0.6$ m.

2039

Set up on the lecture table is a demonstration that shows that it is possible to "measure the wavelength of a laser with a ruler". Make the obvious observations (please do not touch the equipment, the alignment is tricky), and use your data to calculate the wavelength of the laser.

(*Princeton*)

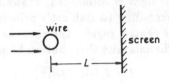

Fig. 2.39

Solution:

As shown in Fig. 2.39, the laser beam falls on a thin wire of diameter d, which has been measured in advance. The diffraction fringes appear on a screen far behind the wire. We obtain the wavelength of the laser by measuring, with the ruler, the distance L between the wire and the screen and the fringe spacing Δ on the screen, and using the relation $\lambda = d\Delta/L$.

2040

A straight horizontal wire of diameter 0.01 mm is placed 30 cm in front of a thin lens with focal length $= +20$ cm.

(a) At what distance from the lens and with what magnification will a sharp image of the wire be focused?

(b) If collimated light is incident from the left, parallel to the optic axis, and a screen is placed at the back focal plane (20 cm behind) of the lens, what diffraction pattern will you observe? Be as quantitative as you can.

(*Princeton*)

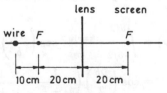

Fig. 2.40

Solution:

(a) The lens formula

$$\frac{1}{u} + \frac{1}{v} = \frac{1}{f}$$

with $u = 30$ cm, $f = 20$ cm yields $v = 60$ cm. The lateral magnification is $M = -\frac{v}{u} = -2$.

(b) As the incident light is collimated, Fraunhofer diffraction by the wire will take place. According to Babinet's principle, the pattern on the screen is, except for the central region, the same as that by a single slit of the same width, with the intensity distribution being given by

$$I = I_0 (\sin \beta / \beta)^2 ,$$

where $\beta = \pi d \sin \theta / \lambda \approx \pi d y / (f \lambda)$, y being the distance on the screen from the optical axis, and d is the diameter of the wire.

2041

Optical apodization is the process by which the aperture function of the system is altered to redistribute the energy in the diffraction pattern. To improve the resolution, consider the example where the transmission is modified by a cosine function, $\cos \pi x / b$, between $-b/2$ and $+b/2$, where b is the slit width.

(a) Calculate the position of the n-th minimum of the single-slit Fraunhofer diffraction pattern, and compare it with the position of the n-th minimum of the unapodized slit of the same width b.

(b) Calculate the intensity which appears in the apodized diffraction pattern halfway between the $(n-1)$-th and n-th minima, which is near its n-th maximum. In each case, normalize your result to the intensity of the first maximum.

(UC, Berkeley)

Solution:

For the apodized aperture, the light amplitude of far-field diffraction is

$$A(\nu) = \int_{-b/2}^{b/2} e^{ikx \sin \theta} \cos \left(\frac{\pi x}{b} \right) dx$$

$$= \cos \left(\frac{\nu b}{2} \right) \left(\frac{1}{\nu - \frac{\pi}{b}} - \frac{1}{\nu + \frac{\pi}{b}} \right) ,$$

or

$$I \sim A^2 = \cos^2\left(\frac{\nu b}{2}\right) \frac{4\left(\frac{\pi}{b}\right)^2}{[\nu^2 - \left(\frac{\pi}{b}\right)^2]^2},$$

where

$$\frac{\nu b}{2} = \frac{\pi b}{\lambda}\sin\theta, \text{ i.e., } \nu = \frac{2\pi}{\lambda}\sin\theta.$$

The position of the n-th minimum is given by

$$\frac{\nu b}{2} = \frac{\pi}{2}(2n + 1)$$

or

$$\sin\theta = \frac{\lambda}{2b}(2n + 1).$$

For the unapodized aperture, the intensity of Fraunhofer diffraction is

$$I \sim b^2 \left(\frac{\sin\frac{\nu b}{2}}{\frac{\nu b}{2}}\right)^2$$

and the position of the n-th minimum is given by $\frac{\nu b}{2} = n\pi$ or

$$\sin\theta' = \frac{n\lambda}{b}.$$

Hence

$$\sin\theta - \sin\theta' = \frac{\lambda}{2b}.$$

(b) The position of the $(n-1)$-th minimum for the apodized aperture is given by

$$\sin\theta = \frac{\lambda}{2b}(2n - 1).$$

The position midway between the $(n-1)$-th and n-th minima is then given by $\sin\theta = n\lambda/b$. It is seen that the intensity is near the n-th maximum given by $\frac{\nu b}{2} = n\pi$. At this position the intensity is

$$I \sim A^2 = \frac{4}{\left(\frac{\pi}{b}\right)^2(4n^2 - 1)^2}.$$

Let the intensity of the first (central) maximum be I_0, which is given by $\frac{\nu b}{2} = 0$. Therefore, for the apodized aperture,

$$\frac{I}{I_0} = \frac{1}{(4n^2 - 1)^2},$$

while for the unapodized aperture, I is given by $\frac{\nu b}{2} = n\pi - \frac{\pi}{2}$, so that

$$\frac{I}{I_0} = \frac{4}{[(2n-1)\pi]^2} \ .$$

Figure 2.41 and Fig. 2.42 show that making use of the apodized aperture has put more energy into the central maximum at the expense of the secondary maxima.

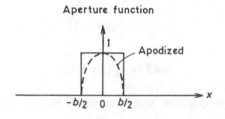

Fig. 2.41

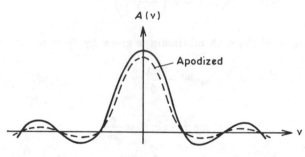

Fig. 2.42

2042

Sunlight enters a large darkroom through a hole 2 cm square, which is covered with a parallel set of wires of width $w/2$, spaced w between centers. The sunlight falls on photographic plates at distances d behind the hole (Fig. 2.43). The sun's angular diameter is about 0.01 radian; describe the predominant character of the images for the followig cases:

$$(w, d) = \text{(a)} \quad (1 \text{ mm}, 1 \text{mm})$$
$$\text{(b)} \quad (1 \text{ mm}, 10 \text{ m})$$
$$\text{(c)} \quad (0.005 \text{ mm}, 10 \text{ m})$$

(*Wisconsin*)

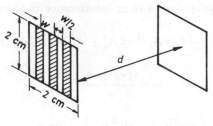

Fig. 2.43

Solution:

(a) As $d = w$, geometric optics applies. There are shadows of the square hole and the wires.

(b) As $w \ll d$, $\alpha = 0.01 \ll 1$, Fraunhofer diffraction pattern occurs on the plate. The grating equation $w \sin \theta \simeq w\theta = m\lambda$, where m is the order of diffraction, gives $\theta \approx \lambda/w = 0.0005 \ll \alpha$ for $\lambda = 5000\,\overset{\circ}{\text{A}}$. Thus the pattern diffracted by the wire set for different colors would overlap and give uniform intensity, i.e., no fringes, on the plate.

(c) For $w = 0.005$ mm, $\theta \approx 0.1 \gg \alpha$. Colorful diffraction patterns are seen on the plate.

2043

(a) Consider the Fraunhofer diffraction pattern due to two unequal slits. Let a and b be the unequal slit widths and c the distance between their centers. Derive an expression for the intensity of the pattern for any diffraction angle θ, assuming the arrangement to be illuminated by perpendicular light of wavelength λ.

(b) Use your formula from (a) to obtain expressions for the pattern in the following special cases and make a sketch of those patterns:

  (1) $a = b$,
  (2) $a = 0$.

(*UC, Berkeley*)

Solution:

(a) The diffraction amplitudes due to slits a and b separately are

$$A_{\mathrm{a}} = \frac{a \sin u}{u}, \quad u = \frac{1}{2} ka \sin \theta = \frac{\pi a \sin \theta}{\lambda},$$

$$A_{\mathrm{b}} = \frac{b \sin v}{v}, \quad v = \frac{1}{2} kb \sin \theta = \frac{\pi b \sin \theta}{\lambda}.$$

The resultant intensity due to their interference is given by

$$I(\theta) = |A_a + A_b|^2 = A_a^2 + A_b^2 + 2A_a A_b \cos w,$$

where

$$w = \frac{2\pi c \sin \theta}{\lambda}.$$

Hence

$$I(\theta) = a^2 \frac{sin^2 u}{u^2} + b^2 \frac{\sin^2 v}{v^2} + 2ab \frac{\sin u}{u} \cdot \frac{\sin v}{v} \cos w.$$

(b) (1) When $a = b$, then $u = v$, and

$$I(\theta) = 2a^2 \frac{\sin^2 v}{v^2} (1 + \cos w) = 4a^2 \frac{\sin^2 v}{v^2} \cos^2 \frac{w}{2}.$$

(2) When $a = 0$,

$$I(\theta) = b^2 \frac{\sin^2 v}{v^2}.$$

2044

Consider a monochromatic beam of light incident upon a single slit. Let the wavelength of the light be λ and the slit width be $w = 5\lambda$ (Fig. 2.44).

(a) Sketch the intensity pattern as a function of angle in the region far from the slit.

(b) Calculate the positions of the first maximum and the first minimum.

(c) It is now desired to uniformly phase-shift the light incident upon the upper (i.e., $y > w/2$) half of the slit by exactly 180°. Design an idealized apparatus to accomplish this and sketch the resulting intensity pattern.

(d) Calculate the intensity vs. angle; give the position of the first minimum and an estimate of that for the first maximum.

(*Chicago*)

Solution:

(a) A Fraunhofer diffraction intensity pattern will be observed on a screen far from the slit as shown in Fig. 2.45 as a function of the angle θ from the axis of symmetry.

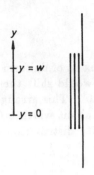

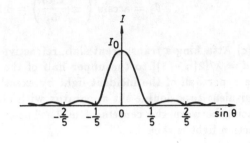

Fig. 2.44 Fig. 2.45

Analytically the intensity distribution is

$$I = I_0 \frac{\sin^2 u}{u^2},$$

where

$$u = \frac{\pi w \sin \theta}{\lambda}.$$

(b) The position of the first minimum is given by

$$\frac{\pi w \sin \theta}{\lambda} = 5\pi \sin \theta = \pm\pi.$$

Hence

$$\theta = \arcsin\left(\pm\frac{1}{5}\right) = \pm 11.5°.$$

The position of the maxima are given by

$$\frac{dI(u)}{du} = 0,$$

or

$$\tan u = u.$$

This is a transcendental equation, the solution of which can be obtained graphically. Next to the central maximum for which $u = 0$ the first maximum is given by

$$u_1 = \pm 1.43\pi.$$

Hence

$$\theta_1 = \arcsin\left(\pm\frac{1.43}{5}\right) = \pm 16.6°.$$

(c) Attaching a transparent slab, refractive index n, of uniform thickness $d = \lambda/[2(n-1)]$ to the upper half of the slit would shift the phase of the upper half of the incident light by exactly 180°. This arrangement is equivalent to a double-slit apparatus with identical slit width $w/2$ and a spacing between the centers of $w/2$. The intensity distribution of the diffraction light is then

$$I = I_0 \frac{\sin^2 u'}{u'^2} \cdot \cos^2 v.$$

where

$$u' = \frac{\pi w \sin\theta}{2\lambda} = \frac{5}{2}\pi\sin\theta,$$

$$v = \frac{\pi w \sin\theta}{2\lambda} + \frac{\pi}{2} = \frac{5}{2}\pi\sin\theta + \frac{\pi}{2}.$$

Hence

$$I = I_0 \sin^4\left(\frac{5}{2}\pi\sin\theta\right)\Bigg/\left(\frac{5}{2}\pi\sin\theta\right)^2.$$

Figure 2.46 shows $\left(\frac{\sin u'}{u'}\right)^2$, $\cos^2 v$ ad I as functions of $\sin\theta$.

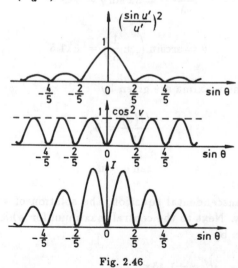

Fig. 2.46

(d) The first minimum occurs at $\sin\theta = 0$, i.e., the center. The next minimum is given by

$$\frac{5}{2}\pi \sin \theta = \pm\pi ,$$

i.e., at

$$\sin \theta = \pm\frac{2}{5} ,$$

or

$$\theta = \arcsin \left(\pm\frac{2}{5} \right) = \pm 23.5° .$$

The position of the first maximum is approximately given by

$$\sin \theta = \pm\frac{1}{5} .$$

Hence

$$\theta = \arcsin \left(\pm\frac{1}{5} \right) = \pm 11.5° .$$

Note that in this case, the central maximum disappears.

2045

Sketch the interference pattern on a screen at the focal point of converging lens which arises from a double slit on which plane monochromatic light is incident. The two identical slits, having width a, are separated by a center-to-center distance d, and $d/a = 5$.

(*Wisconsin*)

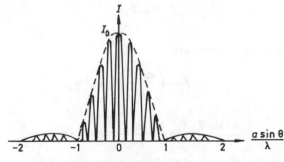

Fig. 2.47

Solution:

The intensity distribution on the screen is given by

$$I = I_0 \left(\frac{\sin \beta}{\beta} \right)^2 \cos^2 \gamma,$$

where

$$\beta = \frac{\pi a \sin \theta}{\lambda}, \gamma = \frac{\pi d \sin \theta}{\lambda} = 5\beta.$$

The interference pattern is shown in Fig. 2.47. Note that as $\frac{d}{a} = 5$ orders $5, 10, 15 \ldots$ are missing.

2046

A plane wave of light from a laser has a wavelength of $6000 \overset{\circ}{A}$. The light is incident on a double slit. After passing through the double slit the light falls on a screen 100 cm beyond the double slit. The intensity distribution of the interference pattern on the screen is shown (Fig. 2.48). What is the width of each slit and what is the separation of the two slits?

(*Wiscconsin*)

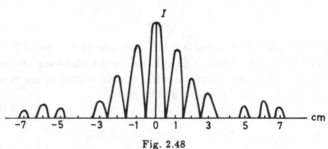

Fig. 2.48

Solution:

For double-slit interference

$$I \sim \left(\frac{\sin \beta}{\beta} \right)^2 \cos^2 \gamma,$$

where

$$\beta = \frac{\pi b \sin \theta}{\lambda}, \gamma = \frac{\pi d \sin \theta}{\lambda}, \text{ with } \sin \theta \approx \frac{y}{D},$$

b being the width of each slit, d the separation of their centers, D the distance between the screen and the double slit, and λ the wavelength of the incident light. The minima are given by $\gamma = \left(n + \frac{1}{2}\right)\pi, n = \pm 1, \pm 2, \ldots$. Thus the fringe spacing Δy is given by

$$\Delta y = \frac{\Delta \gamma}{\pi} \frac{\lambda D}{d} = \frac{\lambda D}{d}.$$

Hence, for $\Delta y = 1$ cm as from Fig. 2.48,

$$d = \lambda D / \Delta y = 6 \times 10^{-3} \text{ cm}.$$

The pattern shows the missing order $d/b = 4$, from which we get the width of each slit, $b = d/4 = 1.5 \times 10^{-3}$ cm.

2047

Shown in Fig. 2.49 is the Fraunhofer diffraction pattern resulting from 3 slits. The slit widths are w, slit separation is d, the distance between the screen and slits is f, and the wavelength of the light is λ. Obtain expressions for x, D, and I_0/I_1 in terms of the parameters of this experiment.

(*Wisconsin*)

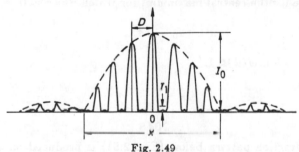

Fig. 2.49

Solution:

The intensity distribution resulting from 3 slits is given by

$$I \sim A_0^2 \left(\frac{\sin \beta}{\beta}\right)^2 \left[\frac{\sin(3\gamma)}{\sin \gamma}\right]^2,$$

where

$$\beta = \frac{\pi w \sin \theta}{\lambda}, \gamma = \frac{\pi d \sin \theta}{\lambda}, \sin \theta \approx \frac{x}{2f}.$$

As $\lim\limits_{\gamma \to m\pi} \frac{\sin(3\gamma)}{\sin \gamma} \to 3, m = 0, \pm 1, \pm 2, \ldots$, the principal maximum occurs when $\theta = 0$. The first maximum occurs when $\frac{\pi d \sin \theta}{\lambda} = \pi$. The fringe spacing is given by $\Delta\gamma = \pi$, or $D \approx f \sin\theta$, i.e.,

$$D \approx \frac{\lambda f}{d}.$$

The first minimum of the diffraction intensity occurs when $\frac{\pi w \sin\theta}{\lambda} = \pi$, or

$$x = \frac{2\lambda f}{w}.$$

The first and second interference minima occur when $\frac{3\pi d \sin\theta}{\lambda} = \pi$ and 2π respectively. The first subsidiary maximum will occur between them, i.e., when

$$3\gamma \approx 3\pi/2, \text{ or } \sin\theta \approx \frac{\lambda}{2d}.$$

Thus gives

$$I_1 \sim A_0^2 \cdot \left(\frac{\sin\frac{\pi w}{2d}}{\frac{\pi w}{2d}}\right)^2 \cdot \left(\frac{\sin\frac{3\pi}{2}}{\sin\frac{\pi}{2}}\right)^2 = A_0^2 \left(\frac{\sin\frac{\pi w}{2d}}{\frac{\pi w}{2d}}\right)^2,$$

We also have for the central maximum, for which $\beta = \gamma = 0$,

$$I_0 \sim 9A_0^2.$$

Hence $I_0/I_1 \approx 9$ if $w/d \ll 1$.

2048

The diffraction pattern below (Fig. 2.51) is produced on a screen by light passing through three distant slits of width w and spacing d (Fig. 2.50). How will the envelope width A and the separation of the peaks B change if:

    (a) The slit width w is increased?
    (b) The slit spacing d is increased?
    (c) The wavelength of the light is increased?

The amplitudes of the E-field arriving at point 2 on the screen are represented on the right by three equal vectors summed with the appropriate phases for this position (Fig. 2.52). As shown, they add to zero, which is why there is zero intensity at this point on the screen. Make a similar vector diagram showing the relative phases:

    (d) at point 1.

(e) at point 3.

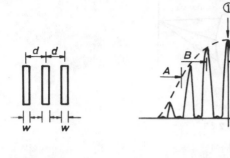

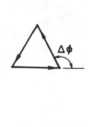

Fig. 2.50   Fig. 2.51   Fig. 2.52

Solution:

(a) The light intensity distribution on the screen is represented by

$$I \propto \left(\frac{\sin \frac{N \pi d \sin \theta}{\lambda}}{\sin \frac{\pi d \sin \theta}{\lambda}} \right)^2 \cdot \left(\frac{\sin \frac{\pi w \sin \theta}{\lambda}}{\frac{\pi w \sin \theta}{\lambda}} \right)^2,$$

where $N = 3$, $\sin \theta \approx x/l$ (l is the focal length of a lens placed behind the slits or the distance between the screen and the slits).

From the formula we get the envelope width $A \propto \lambda l/w$ and the separation of the peaks $B \propto \lambda l/d$. Therefore we conclude:

(a) A decreases and B remains unchanged when w is increased.

(b) A remains unchanged but B decreases when d is increased.

(c) Both A and B will increase when the wavelength of the light is increased.

(d) The vector diagram for point 1 is ⟶⟶⟶

(e) The vector diagram for point 3 is ⇉.

2049

A plane wave of wavelength λ is incident on a system having 3 slits of width a separated by distances d. The middle slit is covered by a filter which introduces a 180° phase change.

Calculate the angle θ for the

(a) first diffraction minimum,

(b) first interference minimum,

(c) first interference maximum.

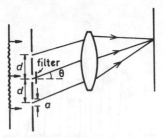

Fig. 2.53

Solution:

By the Huygens-Fresnel principle, the amplitude at a point on the screen is given by (for Fraunhofer diffraction)

$$\psi = \frac{A_0}{a} e^{-iwt} \left[\int_0^a e^{i2\rho x} dx \right.$$

$$\left. + \int_d^{d+a} e^{i(2\rho x + \pi)} dx + \int_{2d}^{2d+a} e^{i2\rho x} dx \right]$$

$$= A_0 e^{-iwt} e^{i\rho a} \frac{\sin(\rho a)}{\rho a} \left(1 - e^{i2\rho d} + e^{i4\rho d} \right) e^{i2\rho d}$$

$$= A_0 e^{i(\beta + 4\gamma - wt)} \frac{\sin \beta}{\beta} \frac{\cos(3\gamma)}{\cos \gamma} ,$$

where $\rho = \frac{\pi}{\lambda} \sin \theta, \beta = \rho a, \gamma = \rho d$. Thus

$$I \sim \left(\frac{\sin \beta}{\beta} \right)^2 \left[\frac{\cos(3\gamma)}{\cos \gamma} \right]^2 .$$

(a) The first diffraction minimum occurs at

$$\beta = \pm \pi , \text{ i.e., } \theta = \pm \sin^{-1} \left(\frac{\lambda}{a} \right) .$$

(b) The first interference minimum occurs at

$$3\gamma = \pm \frac{\pi}{2} , \text{ i.e., } \theta = \pm \sin^{-1} \left(\frac{\lambda}{6d} \right) .$$

(c) The frist interference maximum occurs at

$$\gamma = \pm\frac{\pi}{2}, \text{ i.e., } \theta = \pm\sin^{-1}\left(\frac{\lambda}{2d}\right).$$

2050

Consider an opaque screen with 5 equally spaced narrow slits (spacing d) and with monochromatic plane-wave light (wavelength λ) incident normally. Draw a sketch of the transmitted intensity vs. angle to the normal for $\theta = 0$ to approximately $\theta = 1/5$ radian. Assume $d/\lambda = 10$. Your sketch should show the maxima and minima of intensity. Approximately what is the ratio in intensity of the least intense to the most intense peak? Approximately what is the angular distance of the first intense peak away from $\theta = 0$?

(*Wisconsin*)

Solution:

For multiple-slit interference,

$$I \propto \left(\frac{\sin\frac{N\delta}{2}}{\sin\frac{\delta}{2}}\right)^2 = \left(\frac{\sin\frac{5\pi d\sin\theta}{\lambda}}{\sin\frac{\pi d\sin\theta}{\lambda}}\right)^2 = \left(\frac{\sin(50\pi\sin\theta)}{\sin(10\pi\sin\theta)}\right)^2.$$

For $\theta = 0$ to $\frac{1}{5}$ rad we may take the approximation $\sin\theta \approx \theta$. Thus

$$I \propto \left(\frac{\sin(50\pi\theta)}{\sin(10\pi\theta)}\right)^2.$$

Therefore, intensity maxima occur when $10\pi\theta = m\pi$, i.e., $\theta = m/10$, where $m = 0, \pm1, \pm2, \ldots$. Intensity minima occur when $50\pi\theta = n\pi$, i.e., $\theta = n/50$ where n is an integer such that $n \neq 0, \pm5, \pm10, \ldots$.

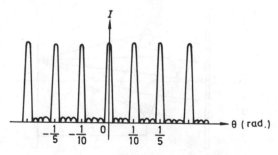

Fig. 2.54

Figure 2.54 shows the intensity distribution. The least intense peak is midway between two adjacent most intense peaks, for which

$$\theta = \frac{1}{2} \times \frac{1}{10} = 0.05 \text{ rad}.$$

The ratio in intensity of the least to the most intense peak is

$$\frac{I(\theta = 0.05)}{I(\theta = 0)} = \left(\frac{\sin(50\pi \times 0.05)}{\sin(10\pi \times 0.05)}\right)^2 \bigg/ \left(\frac{\sin(50\pi \times 0)}{\sin(10\pi \times 0)}\right)^2$$

$$= \left[\frac{\sin(0.5\pi)}{\sin(0.5\pi)}\right]^2 \bigg/ \left[\lim_{\varepsilon \to 0}\frac{\sin(50\pi\varepsilon)}{\sin(10\pi\varepsilon)}\right]^2 = \frac{1}{25}.$$

The sketch shows the angular distance of the first subsidiary peak from $\theta = 0$ to be

$$\frac{3}{2} \times \frac{1}{50} = 0.03 \text{ rad}.$$

2051

A plane monochromatic wave (wavelength λ) is incident on a set of 5 slits spaced at a distance d (Fig. 2.55). You may assume that the width of the individual slits is much less than d. For the resulting interference pattern, which is focused on a screen, compute either analytically or approximately the following. (Hint: The use of phasor diagrams may be helpful.)

(a) Angular width of central image (angle between principal maximum and first minimum.)

(b) Intensity of subsidiary maximum relative to principal maximum intensity.

(c) Angular position of first subsidiary maximum.

(*Wisconsin*)

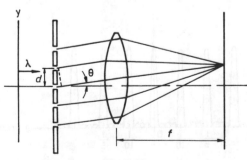

Fig. 2.55

Solution:

Each slit contributes an equal phasor A_0. These contributions can be added up in the phasor diagrams of Fig. 2.56. When $\theta = 0$, all the five phasors are in phase and the resultant phasor is a straight line of $A = 5A_0$, representing the principal maximum (see Fig. 2.56 (a)). For the first minimum, the phasor diagram reduces to a pentagon (see Fig. 2.56 (b)). Here we have $\delta = \frac{2\pi d \sin \theta}{\lambda} = \frac{2}{5}\pi$, i.e., $\theta \approx \sin \theta = \frac{\lambda}{5d}$. For the first

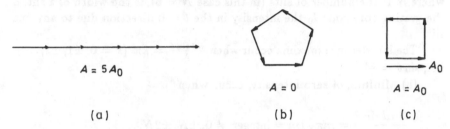

$$A = 5A_0$$

$$(a) \qquad \qquad A = 0 \qquad \qquad A = A_0$$

$$(b) \qquad \qquad (c)$$

Fig. 2.56

subsidiary maximum, the resultant phasor is equal to A_0 (see Fig. 2.56 (c)). Here

$$\delta = \frac{2\pi d \sin \theta}{\lambda} = \frac{\pi}{2}, \text{ i.e., } \theta \approx \sin \theta = \frac{\lambda}{4d}.$$

(a) Angular width of central image is $\theta \approx \frac{\lambda}{5d}$.

(b) Intensity of subsidiary maximum relative to that of the principal maximum is about

$$\left(\frac{A_0}{5A_0}\right)^2 = \frac{1}{25}.$$

(c) Angular position of the first subsidiary maximum is about $\theta \approx \frac{\lambda}{4d}$.

2052

A plane light wave of wavelength λ is incident normally on a grating consisting of 6 identical parallel slits a distance d apart.

(a) At what angles with the normal are the interference maxima?

(b) What is the angular width of the interference maxima, say from the maximum to the nearest null?

(c) What is the ratio of the intensity at an interference maximum to the intensity obtained for a single slit?

(d) Sketch the interference pattern as a function of angle.

(*Wisconsin*)

Solution:

(a) The intensity distribution for Fraunhofer diffraction is given by

$$I = I_0 \frac{\sin^2\left(\frac{\pi w \sin\theta}{\lambda}\right)}{\left(\frac{\pi w \sin\theta}{\lambda}\right)^2} \frac{\sin^2\left(\frac{N\pi d\sin\theta}{\lambda}\right)}{\sin^2\left(\frac{\pi d\sin\theta}{\lambda}\right)} ,$$

where N is the number of slits (in this case $N = 6$), w the width of a slit, d the grating constant, I_0 the intensity in the $\theta = 0$ direction due to any one slit.

The interference maxima occur when $\frac{\pi d\sin\theta}{\lambda} = k\pi, (k = 0, \pm 1, \pm 2, \dots)$, i.e., $\sin\theta = \frac{k\lambda}{d}$.

(b) Minima, of zero intensity, occur when

$$\frac{N\pi d\sin\theta}{\lambda} = m\pi , \quad (m = \text{integer} \neq 0, \pm N, \pm 2N, \dots) ,$$

i.e.,

$$\frac{\pi d\sin\theta}{\lambda} = \pm\pi/N , \pm 2\pi/N, \dots , (N-1)\pi/N, (N+1)\pi/N , \dots .$$

Hence, the angular width of the interference maxima is $\Delta\theta \approx \lambda/(Nd)$.

(c) $I(0) = N^2 I_0 = 36 I_0$.

(d) The intensity distribution as a function of angle is shown in Fig. 2.57.

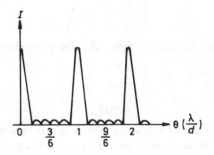

Fig. 2.57

2053

A transmission type diffraction grating having 250 lines/mm is illuminated with visible light at normal incidence to the plane of the grooves.

What wavelengths appear at a diffraction angle of 30°, and what colors are they?

(*Wisconsin*)

Solution:

The grating equation $d \sin \theta = k\lambda$ (k = integer) gives for $d = 1/250$ mm and $\theta = 30°$

$$\lambda = 20000/k \quad \overset{\circ}{\text{A}} .$$

For visible light $(4000\overset{\circ}{\text{A}} - 7000\overset{\circ}{\text{A}})$ we get the following colors,

$$\lambda_1 = 4000\,\overset{\circ}{\text{A}} \quad (k = 5), \text{ violet},$$

$$\lambda_2 = 5000\,\overset{\circ}{\text{A}} \quad (k = 4), \text{ green},$$

$$\lambda_3 = 6670\,\overset{\circ}{\text{A}} \quad (k = 3), \text{ red}.$$

2054

A diffraction grating is ruled with N lines spaced a distance d apart. Monochromatic radiation of wavelength λ is incident normally on the grating from the left; parallel radiation emerges from the grating to the right at an angle θ and is focused by a lens onto a screen.

(a) Derive an expression for the intensity distribution of the interference pattern observed on the screen as a function of θ. Neglect the effect of diffraction due to finite slit width.

(b) The resolving power $\Delta\lambda/\lambda$ of the grating is a measure of the smallest wavelength difference which the grating can resolve. Using the Rayleigh criterion for resolution and the result of part (a), show that the resolving-power in order m is $\frac{\Delta\lambda}{\lambda} = \frac{1}{mN}$.

(*Wisconsin*)

Solution:

(a) The phase difference between rays coming from adjacent slits is $\phi = kd \sin \theta = (2\pi/\lambda)d \sin \theta$. Thus, the total electric field at a point on the screen is a sum of N terms:

$$E = Ae^{-ikr_1}[1 + e^{-i\phi} + e^{-2i\phi} + \ldots + e^{-(N-1)i\phi}]$$

$$= Ae^{-ikr_1} \cdot e^{-i(N-1)\phi/2} \cdot \frac{\sin \frac{N\phi}{2}}{\sin \frac{\phi}{2}} .$$

The intensity of the interference pattern is then

$$I = I_0 \frac{\sin^2\left(\frac{N\pi}{\lambda}d\sin\theta\right)}{\sin^2\left(\frac{\pi}{\lambda}d\sin\theta\right)}.$$

(b) The Rayleigh criterion states that two spectral lines are just resolved when the maximum of one line coincides with the minimum of the other. Thus we require that light of wavelength $\lambda + \Delta\lambda$ forms its principal maximum of order m at the same angle so that for the first minimum of wavelength λ in that order. For order m, the principal maximum is given by $\frac{\phi}{2} = m\pi$ and the first minimum by $\frac{N\phi}{2} = Nm\pi - \pi$. Hence we have

$$\frac{N\pi}{\lambda + \Delta\lambda}d\sin\theta = Nm\pi - \pi, \quad \frac{N\pi}{\lambda}d\sin\theta = Nm\pi.$$

Taking the ratio we have

$$\frac{\lambda + \Delta\lambda}{\lambda} = \frac{Nm}{Nm - 1}, \text{ or } \frac{\Delta\lambda}{\lambda} = \frac{1}{Nm - 1} \approx \frac{1}{Nm}.$$

2055

(a) Calculate the intensity distribution of radiation diffracted from a grating whose rulings' width is negligible.

(b) Find the number of the subsidiary maxima between consecutive principal maxima.

(c) Are the intensities of these subsidiary maxima identical? Explain.

(*Wisconsin*)

Solution:

(a) $I = I_0 \frac{\sin^2 N\varphi/2}{\sin^2 \varphi/2}$, where $\varphi = \frac{2\pi d\sin\theta}{\lambda}$ is the phase difference between two adjacent diffracted beeams of light, d the grating constant, N the total number of the rulings.

(b) $N - 2$.

(c) The intensities of these subsidiary maxima are not identical, since between 0 and $\pi/2$, as $\varphi/2$ increases, the denominator increases while the numerator is a periodic function of $N\varphi/2$. The intensities, being proportional to the quotients, do not remain constant. The intensity contribution for $N = 6$ is shown in Fig. 2.58.

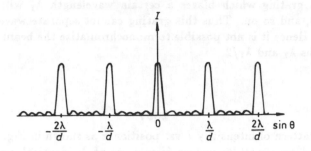

Fig. 2.58

2056

A useful optical grating configuration is to diffract light back along itself.

(a) If there are N grating lines per unit length, what wavelength(s) is (are) diffracted back at the incident angle θ, where θ is the angle between the normal to the grating and the incident direction?

(b) If two wavelengths are incident, namely λ_1 and $\lambda_1/2$, is it possible to monochromatize the beam with this configuration? How?

(*Wisconsin*)

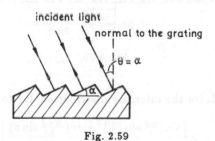

Fig. 2.59

Solution:

(a) This kind of grating is known as blazed grating. As Fig. 2.59 shows, the angle of incidence, θ, is equal to α, which is the angle between the groove plane and the grating plane. Each line has across it an optical path difference $\Delta = 2d\sin\theta$, where $d = 1/N$. Wavelengths for which the grating equation $m\lambda = 2d\sin\theta$ ($m = 1, 2, 3, \ldots$) holds, i.e., $\lambda = 2d\sin\theta/m$, are diffracted back along the incident direction.

(b) A grating which blazes a certain wavelength λ_1 will also blaze $\lambda_1/2, \lambda_1/3$, and so on. Thus this grating cannot separate wavelengths λ_1 and $\lambda_1/2$. Hence it is not possible to monochromatize the beam containing wavelengths λ_1 and $\lambda_1/2$.

2057

The pattern of intensithy I vs. position x as shown in Fig. 2.60 below is measured on a wall 20 meters from a set of N identical, parallel slits. Light of wavelength $\lambda = 6000 \overset{\circ}{A}$ passes through the slits, each of which has width a, and which are spaced from each other by a distance d.

(a) What are the values of N, a and d? (Give a reason for each answer.)

(b) Give an expression for the dashed curve which is the envelope for the large maxima, and explain its physical significance.

(*Wisconsin*)

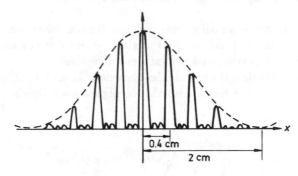

Fig. 2.60

Solution:

(a) The formula for the intensity distribution due to grating diffraction,

$$I = I_0 \left[\frac{\sin\left(\frac{\pi a}{\lambda} \sin \theta\right)}{\frac{\pi a}{\lambda} \sin \theta} \right]^2 \left[\frac{\sin\left(\frac{N\pi d}{\lambda} \sin \theta\right)}{\sin\left(\frac{\pi d}{\lambda} \sin \theta\right)} \right]^2,$$

shows that there are $N - 2$ subsidiary maxima between two adjacent principal maxima. As $N - 2 = 2$ for the above pattern, we have $N = 4$.

The first diffraction null occurs when $\frac{\pi a}{\lambda} \sin \theta = \pi$. We have been given $\sin \theta = (2 \text{ cm})/(20 \text{ m})$, hence

$$a = \frac{\lambda}{\sin \theta} = 0.6 \times 10^{-3} \text{ m}.$$

The first principal maximum occurs when $\frac{\pi d}{\lambda} \sin \theta = \pi$. As we are given $\sin \theta = 0.4 \text{ cm}/20 \text{ m}$, we get

$$d = \frac{\lambda}{\sin \theta} = 3.0 \times 10^{-3} \text{ m}.$$

(b) The dashed curve is given by the single-slit diffraction factor

$$\left[\frac{\sin\left(\frac{\pi a}{\lambda} \sin \theta \right)}{\frac{\pi a}{\lambda} \sin \theta} \right]^2$$

and represents the intensity distribution of light diffracted from each slit. The intensity distribution for N-slit interference, as given by the multi-slit interference factor

$$\left[\frac{\sin\left(\frac{N\pi d}{\lambda} \sin \theta \right)}{\sin\left(\frac{\pi d \sin \theta}{\lambda} \right)} \right]^2,$$

is modulated by the envelope given by the single-slit diffraction intensity distribution.

2058

A laser beam ($\lambda = 6.3 \times 10^{-5}$ cm) is directed at grazing incidence onto a steel machinist's ruler (which is calibrated in 1/16 inch). The light reflects from the surface of the ruler and is projected onto a vertical wall 10 meters away (Fig. 2.61).

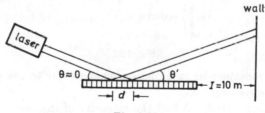

Fig. 2.61

(a) Derive the condition on θ' for interference maxima on the wall. Assume for simplicity that the laser beam is initially almost parallel to the ruler surface.

(b) What is the vertical separation on the wall of the zeroth and first order spots of the pattern?

(*Wisconsin*)

Solution:

(a) The optical path length difference (OPD) of the parallel light rays shown in Fig. 2.61 is

$$\Delta = d(\cos\theta - \cos\theta') \approx d(1 - \cos\theta')$$

as $\theta \approx 0$. Interference maxima occur when $\Delta = m\lambda$ (m is an integer), or

$$\cos\theta' = 1 - m\lambda/d.$$

(b) When $m = 0, \theta' = 0$. When $m = 1, \theta_1' = 1 - \lambda/d = 0.9996$ taking $d = 1$ division ($\frac{1}{16}$ inch). The vertical separation on the wall between the zeroth and first order spots is

$$x = l \cdot \tan\theta_1' = 0.28 \text{ m}.$$

2059

A sound field is created by an arrangement of identical line sources grouped into two identical arrays of N sources each as shown (Fig. 2.62) below.

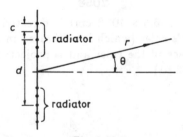

Fig. 2.62

All of the radiators lie in a plane perpendicular to the page, and produce waves of wavelength λ.

(a) Assuming $r \gg d, c, \lambda$ find the intensity of the sound produced as a function of the maximum intensity $I_m, \lambda, \theta, N, c$ and d, the distance between the centers of the arrays.

(b) By taking an appropriate limit, derive an approximate result for the interference pattern produced by two slits of width a whose centers are a distance d apart.

(SUNY, Buffalo)

Solution:

(a) The intensity distribution produced by each array is represented by

$$I \sim \left(\frac{\sin \frac{N\delta}{2}}{\sin \frac{\delta}{2}} \right)^2,$$

where $\delta = 2\pi c \sin \theta / \lambda$. The expression

$$I = I_1 + I_2 + 2\sqrt{I_1 I_2} \cos \varphi$$

for the resultant intensity of two coherent sources of phase difference φ gives

$$I = 2I_1(1 + \cos \varphi)$$
$$= 4I_1 \cos^2(\varphi/2)$$
$$\sim 4 \left(\frac{\sin \frac{N\delta}{2}}{\sin \frac{\delta}{2}} \right)^2 \cos^2 \left(\frac{\pi d \sin \theta}{\lambda} \right),$$

as $\varphi = \frac{2\pi d \sin \theta}{\lambda}$. The maximum intensity is given by $\frac{\delta}{2} = 0$, so that $I_m \sim 4N^2$. Hence

$$I = \frac{I_m}{N^2} \left(\frac{\sin \frac{N\delta}{2}}{\sin \frac{\delta}{2}} \right)^2 \cos^2 \left(\frac{\pi d \sin \theta}{\lambda} \right).$$

(b) For a two-slit system, we take the limit $N \to \infty$ keeping $Nc = a$, a being the width of each slit:

$$\lim_{N \to \infty} I = \lim_{N \to \infty} I_m \frac{\sin^2 \left(\frac{a\pi}{\lambda} \sin \theta \right)}{N^2 \sin^2 \left(\frac{a\pi}{N\lambda} \sin \theta \right)} \cos^2 \left(\frac{\pi}{\lambda} d \sin \theta \right)$$

$$= \lim_{x \to 0} \frac{I_m \sin^2 \left(\frac{a\pi}{\lambda} \sin \theta \right) \cos^2 \left(\frac{\pi d}{\lambda} \sin \theta \right)}{\frac{\sin^2 \left(\frac{a\pi}{\lambda} x \sin \theta \right)}{x^2}}$$

$$= I_m \frac{\sin^2 \left(\frac{a\pi}{\lambda} \sin \theta \right)}{\left(\frac{a\pi}{\lambda} \sin \theta \right)^2} \cdot \cos^2 \left(\frac{\pi}{\lambda} d \sin \theta \right).$$

2060

Figure 2.63 shows a phased array of radar antenna elements spaced at 1/4 wavelength. The transmitting phase is shifted by steps of $\pi/6$ from element to element, i.e., element "0" has 0 phase, element 1 has $\pi/6$ phase shift, element 2 has $2\pi/6$, element 3 has $3\pi/6$, etc.

(a) At what angle θ is the zero order constructive interference transmission lobe?

(b) Are there any secondary lobes?

(*Wisconsin*)

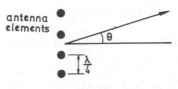

Fig. 2.63

Solution:

Assuming that the complex amplitude of the transmitting wave from the element "0" is A. The complex amplitudes of the successive elements $1, 2, \ldots$ are then respectively represented by $Ae^{i\delta}, Ae^{i2\delta}, \ldots$, where

$$\delta = \frac{2\pi\lambda}{4\lambda}\sin\theta + \frac{\pi}{6} = \frac{\pi}{2}\left(\sin\theta + \frac{1}{3}\right).$$

The resultant amplitude is given by

$$A + Ae^{i\delta} + Ae^{i2\delta} + \ldots + Ae^{i(N-1)\delta} = \frac{Ae^{-\frac{iN\delta}{2}}\sin(N\frac{\delta}{2})}{e^{-\frac{i\delta}{2}}\sin\frac{\delta}{2}},$$

and the resultant intensity is $\sim A^2[\sin(N\delta/2)/\sin(\delta/2)]^2$. Constructive interference transmission lobes occur when $\delta/2 = m\pi, m$ being an integer, or $\frac{\pi}{4}\left(\sin\theta + \frac{1}{3}\right) = m\pi$.

(a) When $m = 0, \sin\theta = -1/3$. Thus the zero order maximum occurs at $\theta = \arcsin(-1/3)$.

(b) For the first secondary maximum, $m = \pm 1$ or $\sin\theta = \pm\frac{11}{3}$. Since $|\sin\theta| < 1$, no secondary lobe can occur.

2061

Describe the Fabry-Perot interferometer. Derive an equation for the positions of the maxima, and indicate their shape. What is the instrument's resolving power?

(*Columbia*)

Solution:

The instrument makes use of the fringes produced in the transmitted light after multiple reflection in the air film between two plane glass plates thinly silvered on the inner surfaces. A lens behind the plates brings the parallel transmitted rays together for interference. Concentric circular rings are formed with the maxima given by

$$2d\cos\theta = m\lambda\,,$$

where d is the separation between the silvered surfaces, θ is the semi-vertex angle subtended by a ring at the center of the lens and m is an integer. For the proof of the above see an optics textbook.

The (chromatic) resolving power of the instrument is the ratio $\frac{\lambda}{\Delta\lambda}$ of two wavelengths λ and $\lambda + \Delta\lambda$ such that the intensity maxima of a given order of the two waves cross at $I = \frac{1}{2}I_{max}$:

$$\frac{\lambda}{\Delta\lambda} = m\frac{\pi r}{1 - r^2}\,,$$

where r^2 is the reflectance of the surfaces.

2062

Find the angular separation in seconds of arc of the closest two stars resolvable by the following reflecting telescope: 8 cm objective, 1.5 meter focal length, 80X eyepiece. Assume a wavelength of 6000 Å. ($1\overset{\circ}{A} = 10^{-8}$ cm).

(Wisconsin)

Solution:

The angular resolving power of the telescope is

$$\Delta\theta_1 = 1.22\frac{\lambda}{D} = 1.22 \times \frac{6000 \times 10^{-8}}{8} \approx 2''\,.$$

The resolving power of the human eye is about 1 minute of arc. Using an eyepiece of magnification 80X, the eye's resolving power is $\Delta\theta_2 = 1'/80 < 2''$. Therefore, the resolvable angular separation of stars using the telescope is $\Delta\theta = \max(\Delta\theta_1, \Delta\theta_2) = 2''$.

2063

You are taking a picture of the earth at night from a satellite. Can you expect to resolve the two headlights of a car 100 km away if you are using a camera which has a 50 mm focal lens with an f number of 2?

(Wisconsin)

Solution:

The Rayleigh criterion gives the angular resolving power as

$$\theta_1 = 1.22 \frac{\lambda}{D} = 1.22 \frac{\lambda}{f'/f}$$

$$\approx 1.22 \times \frac{0.6 \times 10^{-3}}{50/2} \approx 3 \times 10^{-5} \, (\text{rad}),$$

since

$$f = \frac{f'}{D}$$

by definition, where f' is the focal length of the camera lens and D the diameter of its aperture. Taking the separation between the two headlights of a car to be 1 m, their angular separation at the satellite is

$$\theta \approx \frac{1}{100 \times 10^3} \approx 1 \times 10^{-5} \, \text{rad} < \theta_1 \, .$$

Hence, we cannot resolve the two head lights of a car.

2064

By considering the diffraction of light that passes through a single slit of width D, estimate the diameter of the smallest crater on the moon that can be discerned with a telescope if the diameter of the objective lens is 1/2 meter. Assume the wavelength of the light is 5×10^{-7} m and the distance from the earth to the moon is 3.8×10^8 meters.

(Wisconsin)

Solution:

The angular resolving power of the telescope is

$$\theta_1 \approx \lambda/D = 10^{-6} \, \text{rad} \, .$$

The smaller resolvable crater diameter is

$$3.8 \times 10^8 \times 10^{-6} = 380 \, \text{m} \, .$$

2065

The headlights of a car are 1.3 m apart. The pupil of the eye has a diameter of 4 mm. The mean wavelength of light is $\lambda = 5,500 \overset{\circ}{A}$. Estimate the distance at which the headlights can just be resolved.

(*UC, Berkeley*)

Solution:

The angular resolving power of the eye is $\theta = \frac{122\lambda}{D} = 1.68 \times 10^{-4}$ rad. Hence the distance at which the headlights can be resolved is $\frac{1.3}{\theta} = 7.75$ km.

2066

The Space Telescope, as currently planned, will have a mirror diameter of 2 m and will orbit above the earth's atmosphere. Estimate (order of magnitude) the distance between two stars which are 10^{22} cm from the earth and can just be resolved by this telescope.

(*Columbia*)

Solution:

$$\theta \approx \frac{\lambda}{D} = \frac{6 \times 10^{-5}}{2 \times 10^2} = 3 \times 10^{-7} \text{rad} ,$$
$$d \approx 3 \times 10^{-7} \times 10^{22} = 3 \times 10^{15} \text{cm} = 3 \times 10^{10} \text{km} .$$

2067

(a) Show that for Fraunhofer diffraction by a slit the direction of the first minimum on either side of the central maximum is given by $\theta = \lambda/w$, where w is the width of the slit and $w \gg \lambda$.

(b) How wide must a rectangular aperture in front of a telescope be in order to resolve parallel lines one kilometer apart on the surface of the moon? Assume the light has $\lambda = 500 \times 10^{-9}$ m and the distance of the moon is 400,000 kilometers.

(*Wisconsin*)

Solution:

(a) The intensity distribution is given by $I \sim \left(\frac{\sin\beta}{\beta}\right)^2$, where $\beta = \frac{\pi w \sin\theta}{\lambda}$. For the 1st minimum, $\beta = \pi$ and we have $\sin\theta = \frac{\lambda}{w}$, or $\theta \approx \frac{\lambda}{w}$ as $w \gg \lambda$.

(b) $\theta = \frac{1}{4\times10^5}$, $w = \frac{5\times10^{-7}}{\theta} = 0.2$ m.

2068

A 35 mm camera has a lens with a 50 mm focal length. It is used to photograph an object 175 cm in height, such that the image is 30 mm high.

(a) How far from the camera should the person stand?

(b) Estimate the best resolution obtainable on the film using light of $\lambda = 5000\,\overset{\circ}{A}$, if the lens aperture is 1 cm diameter.

(Wisconsin)

Solution:

(a) From

$$\begin{cases} \frac{1}{u} + \frac{1}{v} = \frac{1}{5} \\ m = \frac{3}{175} = \frac{v}{u} \end{cases}$$

we obtain

$$u = 296.7 \text{ cm}, \qquad v = 50.9 \text{ cm}.$$

The person should stand 297 cm away from the camera.

(b) The angular resolving power of a lens is given by

$$\Delta\theta = \frac{1.22\lambda}{D}.$$

As $\Delta\theta = \frac{\Delta x}{v}$, we have

$$\Delta x = \Delta\theta \cdot v = 1.22\lambda v/D = 1.22 \times 5000 \times 10^{-8} \times 51 = 3.1 \times 10^{-3} \text{ cm}.$$

2069

Two stars have an angular separation of 1×10^{-6} radian. They both emit light of wavelengths 5770 and 5790 $\overset{\circ}{A}$.

(a) How large a diameter of the lens in the telescope is needed to separate the images of the two stars?

(b) How large a diffraction grating is needed to separate the two wavelengths present.

Be explicit and complete in your answers, explaining your reasoning.

Solution:

(a) The angular resolving power of a telescope is given by

$$\theta = \frac{1.22\lambda}{D}.$$

For $\lambda = 5790 \overset{\circ}{A}$ we have

$$D = \frac{1.22 \times 5790 \times 10^{-8}}{1 \times 10^{-6}} = 70.6 \text{ cm}.$$

As this is larger than that for $\lambda = 5770 \overset{\circ}{A}$, the diameter needed to separate the two stars should be at least 70.6 cm.

(b) The chromatic resolving power of a grating is given by

$$\frac{\bar{\lambda}}{\Delta\lambda} = mN,$$

where N is the total number of rulings on the grating, $\bar{\lambda}$ the mean wavelength, and m the order of diffraction. One usually uses orders 1 to 3 for observation. For $m = 1$, $N = \frac{\bar{\lambda}}{\Delta\lambda} = \frac{\frac{5770+5790}{2}}{5790-5770} = 289$. For $m = 3, N = 96$.

Hence a diffraction grating of 289 rulings is needed to separate the wavelengths present.

2070

Two parabolic radio antennae, each of diameter D, are separated by a distance $d \gg D$ and point straight up. Each antenna is driven by a transmitter of nominal frequency f, and wavelength $\lambda(= c/f) \ll D$. Be sure to properly label all sketches.

(a) What is the approximate shape of the pattern of radiation at large distance away $(r \gg D^2/\lambda)$ of one antenna if operated alone?

(b) What is the approximate shape of the pattern of radiation at large distances $(r \gg d, r \gg D^2/\lambda)$ if both antennae are driven in phase by the same transmitter?

(c) Suppose the antennae are driven by separate transmitters, each of nominal frequency f. If the pattern of part (b) above is to be maintained essentially stationary for some time t, what is the requirement on the frequency stability of the two transmitters?

(*Wisconsin*)

Solution:

(a) The pattern of radiation at a large distance r for only one antenna is approximately that of Fraunhofer diffraction by a circular aperture:

$$I = I(0) \left[\frac{2J_1(kR \sin \theta)}{kR \sin \theta} \right]^2 ,$$

where $R = D/2, k = 2\pi/\lambda, \sin \theta = \rho/r, \rho$ being the radial distance of the point under consideration from the center of the pattern, and J_1 is Bessel's function of the first kind, order 1.

(b) The waves emitted by the two antennae driven in phase by the same transmitter will interfere with each other and give rise to an intensity distribution

$$I = I(0) \left[\frac{2J_1(kR \sin \theta)}{kR \sin \theta} \right]^2 \cos^2 \alpha ,$$

where $\alpha = kd \sin \theta / 2$.

(c) If the pattern of part (b) above is to be maintained essentially stationary for some time t, the two transmitters should be partially coherent. The time of coherency t_c as determined by

$$t_c \cdot \Delta f \approx 1$$

should be larger than t. Hence we require that

$$\Delta f \approx \frac{1}{t_c} < \frac{1}{t} .$$

2071

A camera (focal length 50 cm, aperture diameter D), sensitive to visible light, is sharply focused on the stars; it is then used without refocusing for an object at a distance of 100 meters. Roughly, what aperture D will give the best resolution for this object? Give D in cm.

(*Wisconsin*)

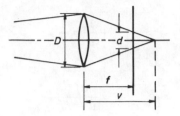

Fig. 2.64

Solution:

Since the camera is sharply focused on stars, the film is located at exactly 50 cm (focal length) from the lens. For an object 100 m away, the image, determined by the lens formula

$$\frac{1}{u} + \frac{1}{v} = \frac{1}{f},$$

lies behind the film. A bright spot of the object at the optical axis will appear on the film as disk of diameter d. The best resolution is obtained if d is of the same size as that of the Airy disk of the lens, which is given by

$$d = \frac{1.22\lambda f}{D}.$$

From the geometry shown in Fig. 2.64 we have

$$\frac{d}{D} = \frac{v - f}{v}.$$

Hence we require

$$d = \frac{(v - f)D}{v} = \frac{fD}{u} = \frac{1.22\lambda f}{D},$$

or $D = \sqrt{1.22\lambda u} \approx 0.82$ cm taking $\lambda = 5500\,\overset{\circ}{\text{A}}$.

2072

An excellent camera lens of 60 mm focal length is accurately focussed for objects at 15 m. For what aperture (stop opening) will diffraction blur of visible light be roughly the same as the defocus blur for a star (at ∞)?

(*Wisconsin*)

Solution:

The lens formula

$$\frac{1}{v} + \frac{1}{u} = \frac{1}{f}$$

gives the image distance as

$$v = \frac{uf}{u - f}.$$

The diffraction blur is given by the Rayleigh criterion $\theta = \frac{1.22\lambda}{D}$. Then the blur diameter is

$$d = \theta v = \frac{1.22\lambda v}{D}.$$

The defocus blur as given by the geometry of Fig. 2.65 is

$$d = \frac{(v - f)D}{f}.$$

If the two are to be equal, we require

$$D = \sqrt{\frac{1.22\lambda v f}{(v - f)}} = \sqrt{1.22\lambda u}.$$

With $\lambda = 5500 \overset{\circ}{A}$ for visible light, $u = 15$ m, we have $D = 0.32$ cm.

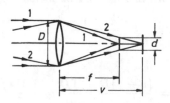

Fig. 2.65

2073

A parabolic mirror of small relative aperture (10 cm dia, 500 cm focal length) is used to photograph stars. Discuss the dominant limitations of resolution, first on axis then off axis. Roughly what is the size of the image "blob" for a star on axis for visible light?

(*Wisconsin*)

Solution:

For photographing stars, the dominant limitation of resolution is the Fraunhofer diffraction by the aperture of the telescope. The diameter of the image "blob" (the Airy disk) of a star on axis is

$$d = \frac{1.22\lambda f}{D}.$$

With $D = 10$ cm, $f = 500$ cm and $\lambda = 5000 \overset{\circ}{A}$ we have

$$d = 0.3 \times 10^{-2} \text{ cm}.$$

For a star off axis, the ray makes an angle θ with the axis of the telescope. So the effective diameter of the aperture reduces to $D\cos\theta$, and the diameter of the Airy disk becomes

$$d' = \frac{1.22\lambda f}{D\cos\theta}.$$

2074

A pinhole camera consists of a box in which an image is formed on the film plane which is a distance P from a pinhole of diameter d. The object is at a distance L from the pinhole, and light of wavelength λ is used (Fig. 2.66).

(a) Approximately what diameter d of the pinhole will give the best image resolution?

(b) Using the pinhole from part (a), approximately what is the minimum distance D between two points on the object which can be resolved in the image?

(*Wisconsin*)

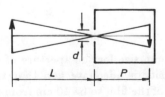

Fig. 2.66

Solution:

(a) By geometrical optics we see that a point on the object would cast a bright disk on the film of diameter Δ_1 given by

$$\frac{\Delta_1}{L+P} = \frac{d}{L}.$$

On the other hand, because of diffraction by the pinhole the point would form a bright Airy disk on the film of diameter

$$\Delta_2 \approx \frac{\lambda P}{d}.$$

The resultant diameter of the image of the point is then

$$\Delta = \Delta_1 + \Delta_2 = \frac{(L+P)d}{L} + \frac{\lambda P}{d}.$$

Minimizing Δ with respect to d gives

$$d = \sqrt{\frac{\lambda L P}{L+P}}.$$

(b) For the images of two close points of separations on the object to be resolved the distance between the two images should not be smaller than Δ_2. Geometric optics gives

$$\frac{D}{L} = \frac{\Delta_2}{P}.$$

Hence $D = \frac{\lambda L}{d} = \sqrt{\frac{\lambda L(L+P)}{P}}$.

2075

Estimate the optimum size for the aperture of a pinhole camera. You may assume that intensity is adequate and that the object viewed is at infinity. Take the plane of the film to be 10 cm from the pinhole (Fig. 2.67).

(*MIT*)

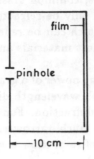

film

pinhole

←——10 cm ——→

Fig. 2.67

Solution:

According to the solution of problem 2074, the optimal size of the aperture is

$$d = \sqrt{\frac{\lambda LP}{L + P}} \approx \sqrt{\lambda P}$$

for

$$L \gg P.$$

Hence $d = 2.3 \times 10^{-2}$ cm taking $\lambda = 5500\,\overset{\circ}{A}$.

2076

Discuss the difference between optics as performed at wavelengths near $100\,\overset{\circ}{A}$ with optics near $5000\,\overset{\circ}{A}$. Specifically, contrast:

(a) Use of lenses.

(b) Use of mirrors.

(c) Chromatic resolving power of gratings of fixed width.

(d) Minimum resolvable angular separation of an image-forming system of fixed diameter.

(*Wisconsin*)

Solution:

(a) Light of wavelenghts near $5000\,\overset{\circ}{A}$ is visible, while that near $100\,\overset{\circ}{A}$, being in the X-ray range, cannot be seen by the eye.

As the refractive indices of glass are greater than unity for visible light and less than unity for X-rays, common glass lenses can be used to focus visible light, while X-rays can only be refracted by crystals.

(b) Unlike visible light which can be reflected by common mirrors, X-rays can penetrate through most materials and can be reflected only when incident at critical angles.

(c) The chromatic resolving power of a grating is given by $\lambda/\Delta\lambda = mN$, where $\Delta\lambda$ is the least resolvable wavelength difference, N the total number of rulings and m the order of diffraction. For given m and N, as λ is much smaller for X-rays, $\Delta\lambda$ is also much narrower.

(d) The minimum angle of resolution is given by

$$\Delta\theta = 1.22\lambda/D\,,$$

where D is the diameter of the aperture. As $\Delta\theta \propto \lambda$, the minimum resolvable angular separation should be much smaller for X-rays.

2077

Compare the merits of the following three instruments for studying atomic spectra (in the visible range).

(1) Plane diffraction grating.

(2) 60° glass prism.

(3) Fabry-Perot interferometer.

Consider as quantitatively as possible such features as resolution, dispersion, range, nature of light-source, etc. Indicate clearly the important physical and geometrical features in each case. (For the sake of definiteness, you may assume all three instruments to have similar 'apertures'.)

(*Wisconsin*)

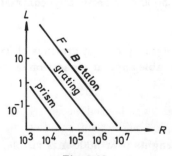

Fig. 2.68

Solution:

For discriminating wavelengths, the glass prism has the lowest resolving power, while the Fabry-Perot interferometer has the highest and can be used to study the fine structure of spectra. The chromatic resolving power of a spectrometer is defined as $R = \frac{\lambda}{\Delta\lambda}$. For a prism, $R = B\left(\frac{dn}{d\lambda}\right)$, where B is the length of the base and n the refractive index for wavelength λ. For a plane grating, $R = mN$, where m is the order of diffraction and N the total number of rulings. For the Fabry-Perot, $R = \frac{m\pi}{1-r}\sqrt{r}$, where r is the reflectance of the mirrors and m the order of interference $\approx \frac{2s}{\lambda}$, s being the plate separation. For example for a wavelength of 500 nm, a flint glass prism $\left(n = 1.6, \frac{dn}{d\lambda} \approx 8 \times 10^2 \text{ cm}^{-1}\right)$ with $B = 10$ cm has $R \sim 10^4$, a plane grating of grating space 14000 lines per inch and width 6 inches has $R \sim 10^5 - 10^6$, a Fabry-Perto has $R \sim 10^6 - 10^7$.

The angular dispersion, defined as $\frac{\Delta\theta}{\Delta\lambda}$, is given by $\frac{2\sin(\frac{\alpha}{2})}{\sqrt{1-n^2\sin^2(\frac{\alpha}{2})}}\left(\frac{dn}{d\lambda}\right)$ for a prism of refracting angle α at minimum deviation, by $\frac{m}{d\cos\theta}$ for a plane grating of grating space d, and by $\frac{\cos\theta}{\lambda}$ for the Fabry-Perot interferometer. For the above instruments the angular dispersions are $1.3 \times 10^3, 10^4$ and $2 \times 10^4 \text{ cm}^{-1}$ respectively.

For a prism the range of wavelengths to be studied is limited only by the absorption of the material. For gratings it is limited to avoid overlapping of fringes of different wavelengths. If a range from λ_1 to λ_2 is observed in the m-th and $(m+1)$-th orders, there is an overlap if $m\lambda_2 < (m+1)\lambda_1$. Consider the limiting case for a Fabry-Perot where the $(m+1)$-th order of λ coincides with the m-th order of $\lambda+\Delta\lambda$. The value $\Delta\lambda = \frac{\lambda}{m} = \frac{\lambda^2}{2d\cos\theta} \approx \frac{\lambda^2}{2d}$ for $\theta \approx 0$ is the largest wavelength range with which the instrument should be illuminated to avoid overlapping of the rings of different colors and is known as the free spectral range.

The light-collecting power L of an instrument depends on its total transmittance T, aperture A and the solid angle Ω subtended by the light gate through $L \approx TA\Omega$. Figure 2.68 shows the limiting R and L for the three types of instrument. The efficiency of an instrument is often expressed as $E = RL$.

2078

A partially elliptically polarized beam of light, propagating in the z direction, passes through a perfect linear polarization analyzer. When the transmission axis of the analyzer is along the x direction, the transmitted intensity is maximum and has the value $1.5 I_0$. When the transmission axis is along the y direction, the transmitted intensity is minimum and has the value I_0.

(a) What is the intensity when the transmission axis makes angle θ with the x-axis? Does your answer depend on what fraction of the light is unpolarized?

(b) The original beam is made to pass first through a quarter-wave plate and then through the linear polarization analyzer. The quarter-wave plate has its axes lined up with the x and y axes. It is now found that the maximum intensity is transmitted through the two devices when the analyzer transmission axis makes an angle of 30° with the x-axis.

Determine what this maximum intensity is and determine the fraction of the incident intensity which is unpolarized.

$(CUSPEA)$

Solution:

(a) Let the intensity of the unpolarized light after passing through the analyzer be I_u. This will remain constant whatever angle the transmission axis makes with the x-axis. Let the intensities of the x and y components of the elliptically polarized light be I_{ex} and I_{ey} respectively. We have

$$I_x = 1.5I_0 = I_u + I_{ex},$$
$$I_y = 1.0I_0 = I_u + I_{ey}.$$

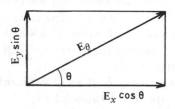

Fig. 2.69

An elliptical vibration can be considered as made up of two mutually perpendicular linear ones 90° out of phase as shown in Fig. 2.69. Its electric field vector can be represented by

$$\mathbf{E}_\theta = \mathbf{E}_x \cos\theta + \mathbf{E}_y \sin\theta.$$

The intensity of the polarized component I_e is then $I_e \sim |\mathbf{E}_0|^2$, or

$$I_e = I_{ex}\cos^2\theta + I_{ey}\sin^2\theta.$$

Hence, writing $I_u = I_u \cos^2 \theta + I_u \sin^2 \theta$, we have

$$I(\theta) = I_e + I_u = (I_{ex} + I_u) \cos^2 \theta + (I_{ey} + I_u) \sin^2 \theta$$
$$= 1.5 I_0 \cos^2 \theta + I_0 \sin^2 \theta .$$

Thus $I(\theta)$ does not depend on what fraction of the light is unpolarized.

(b) A $\frac{\lambda}{4}$ plate introduces a 90° phase change between the two mutually perpendicular components and makes the elliptically polarized light linear polarized. As

$$\frac{\sqrt{I_{ey}}}{\sqrt{I_{ex}}} = \tan 30° = \frac{1}{\sqrt{3}} ,$$

we have

$$I_{ey} = \frac{I_{ex}}{3} .$$

Combining it with the two original equations yields

$$I_{ex} = 0.75 I_0 , \quad I_{ey} = 0.25 I_0 ,$$
$$I_u = 0.75 I_0 .$$

Hence the maximum intensity is

$$I_{ex} + I_{ey} + I_u = 1.75 I_0$$

when $\theta = 30°$. The unpolarized fraction of the incident intensity is

$$\frac{2 I_u}{1.5 I_0 + I_0} = 0.60 .$$

2079

Four perfect polarizing plates are stacked so that the axis of each is turned 30° clockwise with respect to the preceding plate. How much of the intensity of an incident unpolarized beam of light is transmitted by the stack?

(*Wisconsin*)

Solution:

An unpolarized light can be regarded as the resultant of two non-coherent linearly polarized components with mutually perpendicular planes of polarization. Then if the intensity of the incident light is I_0, the intensity of the light emerging from the first polarizing plate is $I_0/2$. According to Malus' law, the intensity of light emerging from a polarizing plate reduces to $\cos^2\theta$ of that of the incident linearly polarized light, where θ is the angle between the transmission axis of the plate and the plane of polarization of the incident light. So the intensity of the light emerging from the fourth plate is

$$I_4 = I_3\cos^2\theta = I_2\cos^4\theta = I_1\cos^6\theta = \frac{I_0\cos^6\theta}{2} = \frac{27I_0}{128}.$$

2080

A thin, quartz crystal of thickness t has been cut so that the optic axis (o.a.) is parallel to the surface of the crystal (Fig. 2.70). For sodium light ($\lambda = 589\,\text{nm}$), the index of refraction of the crystal is 1.55 for light polarized parallel to the optic axis and 1.54 for light polarized perpendicular to the optic axis.

(a) If two beams of light, polarized parallel and perpendicular to the optic axis respectively, start through the crystal in phase, how thick must the crystal be in order that the beams emerge 90° out of phase?

(b) State precisely how to use the above crystal to produce a beam of circularly polarized light.

(*Wisconsin*)

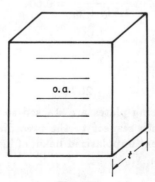

Fig. 2.70

Solution:

(a) $\Delta\phi = \frac{2\pi}{\lambda}t(n_e - n_o) = \frac{\pi}{2}$, giving

$$t = \frac{\lambda}{4(n_e - n_o)} = 14.7\,\mu m \,.$$

(b) Let a beam of unpolarized light pass through a polarizing plate to become plane polarized. The polarized light is allowed to fall normally on the above crystal. If the transmission axis of the polarizing plate makes an angle of 45° with the optic axis of the crystal, the emerging light will be circularly polarized.

2081

Suppose you have been supplied with a number of sheets of two types of optically active material. Sheets of type P are perfect polarizers: they transmit (normally incident) light polarized parallel to some axis $\hat{n}$ and absorb light polarized perpendicular to $\hat{n}$. Sheets of type Q are quarter-wave plates: they advance the phase of (normally incident) light polarized parallel to some axis $\hat{m}$ by $\pi/2$ relative to light polarized perpendicular to $\hat{m}$.

(a) Describe how to combine these materials to produce a device which converts incident light polarized parallel to some axis $\hat{x}$ to outgoing light polarized perpendicular to $\hat{x}$ (see Fig. 2.71). What is the loss in intensity in the device you have constructed?

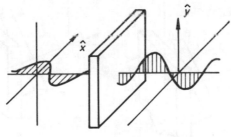

Fig. 2.71

(b) Describe how to combine these materials to produce a "circular polarizer" — a device which converts unpolarized light entering side A into circularly polarized light leaving side B (see Fig. 2.72). What is the minimum possible loss in intensity in this process?

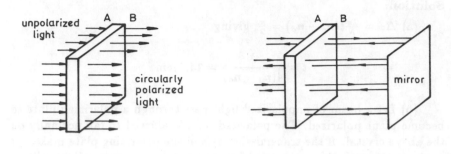

Fig. 2.72   Fig. 2.73

(c) If the power entering side A is I_0 with wavelength λ, what is the magnitude of the torque on the circular polarizer in part (b)?

(d) If the light emerging from the circular polarizer is normally incident on a mirror, what is the loss of intensity of the light in passing through the polarizer a second time? (See Fig. 2.73)

(e) Take two sheets of the circular polarizer you constructed in part (b). Flip one over and put them together as shown in Fig. 2.74. Suppose unpolarized light is incident normally on this system. What is the intensity of transmitted light as a function of θ?

(f) Now reverse the order (see Fig. 2.75). What is the intensity of transmitted light as a function of θ?

(*MIT*)

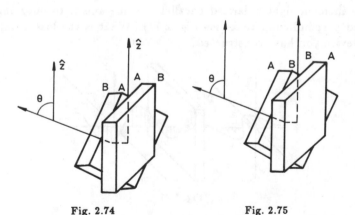

Fig. 2.74   Fig. 2.75

Solution:

(a) Place two P-type sheets together such that the first sheet has its $\hat{n}$ axis at $45°$ to the $\hat{x}$ axis and the second has its $\hat{n}$ axis at $90°$ to the $\hat{x}$

axis. The combined sheet will convert an incident light polarized parallel to $\hat{x}$ into an outgoing light polarized perpendicular to $\hat{x}$. The outgoing intensity is reduced to $I_0 \cos^4 45° = I_0/4$, where I_0 is the incident intensity, according to the law of Malus.

(b) Combine a polarizer (A side) and a $\frac{\lambda}{4}$ wave plate (B side) so that the axes $\hat{n}$ and $\hat{m}$ are orientated at $45°$ to each other. The combined sheet will convert unpolarized light into circularly polarized light. The outgoing intensity is reduced to $I_0/2$ if absorption by the $\frac{\lambda}{4}$ wave plate is negligible.

(c) The angular momentum J and energy W of a circularly polarized light are related by $J = W\omega^{-1}$, where $\omega = 2\pi\nu$, ν being the frequency of the light. The torque on the circular polarizer is

$$\tau = \frac{dJ}{dt} = \omega^{-1}\frac{dW}{dt} = \frac{1}{2}I_0\omega^{-1} = \frac{I_0\lambda}{4\pi c},$$

since only $I_0/2$ goes into the circularly polarized light.

(d) The circularly polarized light is reflected back through the $\frac{\lambda}{4}$ wave plate; the result is a plane polarized light, polarized at right angle to the transmission axis of the polarizer (A side). Therefore, no light is transmitted backwards through the combination.

(e) The intensity of the transmitted light is

$$I = \frac{1}{2}I_0 \cos^2 \theta.$$

(f) The intensity of the transmitted light is

$$I = \frac{1}{2}I_0 \sin^2 \theta.$$

2082

Consider the Babinet Compensator shown in the figure (Fig. 2.76). The device is constructed from two pieces of uniaxial optical material with indices n_e and n_o for light polarized perpendicular and parallel to the optic axis, respectively. A narrow beam of light of vacuum wavelength λ is linearly polarized in the XZ plane at $45°$ to X and Z and propagates through the compensator from left to right along the $+Y$ axis as shown (Fig. 2.77).

(a) For $d \ll L$, calculate the relative phase shift of the X and Z polarized components of the exit beam. Express your answer in terms of $n_o, n_e, \lambda, L, d, x$.

(b) For what values of x will the emerging light be
   (1) linearly polarized?
   (2) circularly polarized?

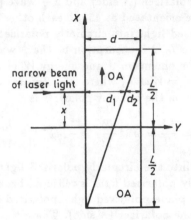

↑ OA= Optic axis in the plane of the paper and parallel to X-axis
• OA= Optic axis perpendicular to the plane of the paper and parallel
    to Z-axis

$d \ll L.$

Fig. 2.76

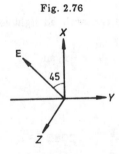

Fig. 2.77

Solution:

(a) As the incident light is polarized at 45° to X- and Z-axes, it is equivalent to two components, polarized in X and Z directions, of amplitudes

$$E_x = E_z = E/\sqrt{2}.$$

The relative phase shifts of the two components of the beam resulting from traversing the left prism and the right prism are respectively given by

$$\Delta\varphi_1 = 2\pi d_1(n_o - n_e)/\lambda$$

and

$$\Delta\varphi_2 = 2\pi d_2(n_e - n_o)/\lambda,$$

where d_1 and d_2 are the distances traveled inside the left prism and the right prism respectively, since the first prism has refractive indices n_o and n_e for the X and Z polarized components respectively and the reverse is true for the second prism.

The relative phase shift of the X and Z polarized components of the exit beam is then

$$\Delta\varphi = \Delta\varphi_1 + \Delta\varphi_2 = 2\pi(n_o - n_e)(d_1 - d_2)/\lambda.$$

By similar triangles we have

$$\frac{d_1}{d} = \frac{\frac{L}{2} + x}{L}.$$

Therefore, as $d = d_1 + d_2$,

$$\Delta\varphi = \frac{4\pi}{\lambda}(n_o - n_e)\frac{xd}{L}.$$

(b) The emerging light would be
   (i) linearly polarized if

$$\Delta\varphi = N\pi, \text{ i.e., } x = \frac{N\lambda L}{4(n_o - n_e)d},$$

   (ii) circularly polarized if

$$\Delta\varphi = \frac{(2N + 1)\pi}{2}, \text{ i.e., } x = \frac{(2N + 1)}{8}\frac{\lambda L}{(n_o - n_e)d},$$

where $N = 0$ or integer.

2083

Consider the modified Young's double-slit arrangement (Fig. 2.78): Q is a monochromatic point source of light with wavelength λ. S_1 is a screen having a long narrow slit and S_2 has two slits of width a separated by a distance $d \gg a$, P_1, P_2, P_3 and P_4 are polarizing filters. For each of the following arrangments, describe and briefly explain the intensity pattern on the screen S_3.

(a) All polarizers removed. (Here derive a formula for the intensity pattern on S_3.)

(b) P_1 removed, P_2 and P_3 have mutually perpendicular pass axes while the axis of P_4 is at 45° to that of P_2.

(c) As in (b) but with P_4 removed.

(d) P_1 in place and at 45° to P_2, P_2 and P_3 still crossed, P_4 perpendicular to P_1.

(CUSPEA)

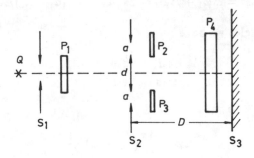

Fig. 2.78

Solution:

(a) The intensity distribution on S_3 is that of the interference fringes produced by a double-slit modulated by the single-slit diffraction factor and is given by

$$I(\theta) \propto \cos^2\left(\frac{1}{2}kd\sin\theta\right) \frac{\sin^2\left(\frac{1}{2}ka\sin\theta\right)}{\sin^2\theta},$$

where $k = 2\pi/\lambda$ and θ is the angle between the diffracted direction and the axis.

(b) As the incoming light on S_2 is unpolarized, the beams leaving P_2 and P_3 (P_2, P_3 crossed) are incoherent and will not interfere whether passing through P_4 or not. We conclude that no interference pattern occurs for (b) and (c). Only the single-slit diffraction images of the slits are seen.

(d) As the incoming light in S_4 is linearly polarized, the beams emerging from P_2 and P_3 are coherent and their components along the axis of P_4

will interfere though the relative phase shift between the two components is π. We get the interference pattern again but the contrast is reversed as compared with that for (a).

2084

A source of circularly polarized light illuminates a screen with two equidistant narrow slits separated by a distance L (see Fig. 2.79). A thin birefringent slab of thickness d is placed over one of the slits. Its index of refraction is n_o parallel to the slit and n_e perpendicular to the slit. A second screen, parallel to the first, is placed at a distance $D \gg L$ behind the first screen. If one of the slits is covered, the total intensity on the second screen is I_0 (neglect diffraction effects). With both slits open, what is the interference pattern that would be observed by someone who measures the total intensity, summed over polarizations?

(*UC, Berkeley*)

Solution:

Passing through the birefringent slab behind S_1 causes a relative phase shift between the two orthogonal components (O and E beams) of

$$\delta = 2\pi d(n_o - n_e)/\lambda,$$

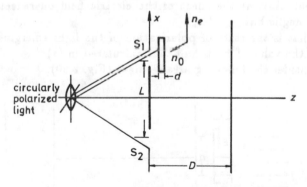

Fig. 2.79

while the phases of the two orthogonal components of light emerging from S_2 remain unchanged. Consider the O and E components arriving at the second screen after leaving S_1 and S_2. The optical path difference between the beams leaving S_1 and S_2 for the two components are

$$\Delta_o = (n_o - 1)d$$

and

$$\Delta_e = (n_e - 1)d$$

respectively.

It is obvious that only when $(n_o - n_e)d = m\lambda$ (m being an integer) can the O and E beams be perfectly coherent at the same time. The contrast of the fringes will then be unity. Otherwise, the contrast is less than unity, especially when $(n_o - n_e)d = (m + 1/2)\lambda$, for which interference fringes disappear.

2085

A beam of light, with $\lambda = 5000\,\overset{\circ}{A}$, traveling in the z-direction is polarized at 45° to the x-direction. It passes through a Kerr cell, i.e., a substance such that $n_x - n_y = KE^2$, where n_x and n_y are the refractive indices for light polarized in the x and y directions. E is the strength of an externally applied electric field in the x-direction. The cell has length 1 cm and $K = 2.5 \times 10^{-6}$ (meter)^2/(volts)^2.

(a) If the light, after passing through the Kerr cell, passes through a polarization analyzer whose plane of polarization is perpendicular to that of the original beam, calculate the smallest value of E which gives maximum transmission. Assume the effect of the electric field on reflection at the Kerr cell is negligible.

(b) What is the state of polarization of the light emerging from the Kerr cell if the value of E^2 is half that calculated in (a)?

(c) Consider the following arrangement (Fig. 2.80).

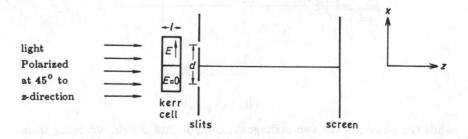

Fig. 2.80

The electric field is applied to the upper half of the Kerr cell only, and, after passing through the Kerr cell, the light passes through two slits as shown. There is no polarization analyzer after the slits. Assuming that the electric field affects n_x but not n_y, discuss the interference pattern, at a large distance beyond the slits, for various values of E^2.

(*UC, Berkeley*)

Solution:

(a) The Kerr cell induces birefringence. As the direction of the electric field is 45° to the direction of polarization of the light, the amplitudes of the components polarized parallel to the x and y directions are the same. When the phase difference δ between the components is

$$\delta = \frac{2\pi l(n_x - n_y)}{\lambda} = \frac{2\pi l K E^2}{\lambda} = (2j + 1)\pi, \; j = 0 \text{ or integer},$$

the Kerr cell acts as a half-wave plate and rotates the plane of polarization by 90° so that it coincides with the transmission axis of the analyzer and gives maximum transmission. Putting $j = 0$ yields the smallest $E, E_{\min}$, which is given by

$$E^2_{\min} = \frac{\lambda}{2Kl} = 10 \quad \text{V}^2/\text{m}^2 \text{ for } l = 1 \text{ cm}.$$

(b) If $E^2 = \frac{1}{2}E^2_{\min} = \frac{\lambda}{4Kl}$, $\delta = \frac{\pi}{2}$ and the emergent light is circularly polarized.

(c) The interference pattern on the far screen is the same as that for a double-slit system, but the contrast is dependent on the polarization states of the light beams emerging from the upper and lower slits. The polarization state of light emerging from the lower slit remains unchanged, while that emerging from the upper slit depends on the strength of the applied electric field. When $\delta = 2j\pi$, where j is an integer, or

$$E^2 = \frac{j\lambda}{Kl},$$

the polarization states of the light beams emerging from the two slits become identical. This gives rise to a perfect interference pattern with the contrast being unity. When

$$\delta = (2j + 1)\pi, \text{ or } E^2 = \frac{(2j + 1)\lambda}{2Kl},$$

the polarization plane of the light emerging from the upper slit is perpendicular to that from the lower slit. There will now be no interference between the beams. In the intermediate cases, the contrast is between unity and zero.

2086

The plane of polarization of plane polarized light transmitted along the optic axis of certain crystals (e.g., quartz or sugar crystals) is rotated by an amount which depends on the thickness of the material. The indices of refraction n_R and n_L for right- and left-circularly polarized light are slightly different in these crystals.

(a) What is this phenomenon called? Describe briefly how the rotation is explained.

(b) Derive an expression for the angle ϕ by which the plane of polarization of light of frequency ω is rotated in passing through a thickness d of the material.

(*Wisconsin*)

Solution:

(a) This phenomenon is called optical activity. An optically active material possesses two indices of refraction, one, n_R, for right-handed rotation and the other, n_L, for left-handed rotation. Since the incident linearly polarized wave can be represented as a superposition of R- and L-circularly polarized waves, in traversing an optically active specimen, the relative phase between the two components will change and the resultant linearly polarized light will appear to have rotated in its plane of polarization.

(b) If $n_R \neq n_L$, the angle of rotation of the L-component after passing through a thickness d of the material is

$$\phi_L = \frac{\omega d}{c} n_L ,$$

and the angle of rotation of the R-component is $\phi_R = \frac{\omega d}{c} n_R$, since the speeds of propagation are c/n_L and c/n_R respectively. The angle ϕ by which the plane of polarization of the resultant light is rotated is then (see Fig. 2.81)

$$\phi = \frac{\phi_R - \phi_L}{2} = \frac{\omega d}{2c} (n_R - n_L) .$$

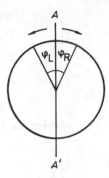

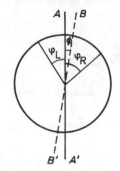

| (a) Before entering | (b) After traversing thickness |
| the material | d of the material. |

Fig. 2.81

2087

Given the following arrangement (Fig. 2.82), where the crystal is assumed to have parallel planes of atoms spaced by distance d. The X-ray tube has anode-cathode potential difference V volts. The angle θ is variable. What is the angle θ_m below which no X-ray intensity will be recorded by the film?

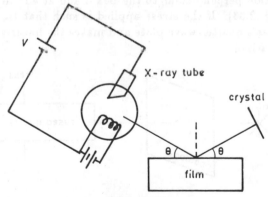

Fig. 2.82

Explain semiquantitatively how one could produce circularly polarized light from an unpolarized source, using a linear polarizer and a slab of fused (i.e., isotropic) quartz and a compressor clamp.

Solution:

(a) When Bragg's law

$$2d \sin \theta = m\lambda \qquad (m = 1, 2, 3, \dots)$$

is satisfied, we get intense spots on the film. The wavelength λ is given by

$$\lambda = \sqrt{150/V} \ \overset{\circ}{\mathrm{A}}.$$

The smallest θ for film recording is given by $m = 1$, for which

$$\theta_{\min} = \arcsin\left(\frac{\lambda}{2d}\right) = \arcsin\left(\frac{1}{2d}\sqrt{\frac{150}{V}}\right).$$

(b) Some stress-sensitive optical materials, such as glass, plastic and melted-quartz, can be made optically anisotropic by the application of mechanical stress. This is known as photoelasticity. The effective optic axis is in the direction of the stress and the induced birefringence $(n_e - n_o)l$, where l is the thickness along the direction of stress, is proportional to the stress. A beam of unpolarized light passing through a polarizer becomes linearly polarized. It falls normally on a slab of melted quartz being compressed along a direction perpendicular to the beam and at 45° to the axis of the polarizer (Fig. 2.83). If the stress applied is such that $(n_o - n_e)l = \lambda/4$, the slab acts as a quarter-wave plate and makes the linearly polarized light circularly polarized.

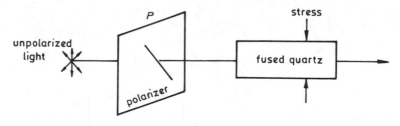

Fig. 2.83

2088

Suppose right-handed circularly polarized light (defined to be clockwise as the observer looks toward the oncoming wave) is incident on an absorbing slab. The slab is suspended by a vertical thread. The light is directed upwards and hits the underside of the slab.

(a) If the circularly polarized light is 1 watt of wavelength 6200 Å, and if all of this light is absorbed by the slab, what is the torque, τ_0, exerted on the slab in dyne-cm?

(b) Suppose that instead of an absorbing slab you use an ordinary silvered mirror surface, so that the light is reflected back at 180° to its original direction. What is the torque now in units of τ_0?

(c) Suppose that the slab is a transparent half-wave plate. The light goes through the plate and does not hit anything else. What is the torque in units of τ_0?

(d) If the slab is a transparent half-wave plate with the top surface silvered, what is the torque in units of τ_0?

(*Chicago*)

Solution:

(a) If a beam of circularly polarized light is completely absorbed by an object on which it falls, an angular momentum

$$\tau_0 = W/\omega$$

is transferred to the object so that a torque is exerted on the object, where W is the amount of energy absorbed per second and ω is the angular frequency of the light. Thus

$$\tau_0 = W/\omega = \frac{W\lambda}{2\pi c} = \frac{1 \times 0.62 \times 10^{-6}}{2\pi \times 3 \times 10^8} = 3.3 \times 10^{-16}\,\text{N.m}$$
$$= 3.3 \times 10^{-9}\,\text{dyn. cm}.$$

(b) The light is reflected back totally such that the direction of rotation of the plane of polarization in space remains unchanged, so that the torque is zero.

(c) After passing through a half-wave plate, the right-handed circularly polarized light becomes left-handed polarized. The torque exerted is equal to the change in angular momentum, which is $2\tau_0$.

(d) The light passes through the half-wave plate twice and remains in the same state as it re-emerges. Hence no torque is exerted on the slab.

2089

A special light source directs 1 watt of light into a flat black (100% absorbing) disc, which is mounted on an axle parallel to the beam (Fig. 2.84). The target starts to spin as it absorbs light.

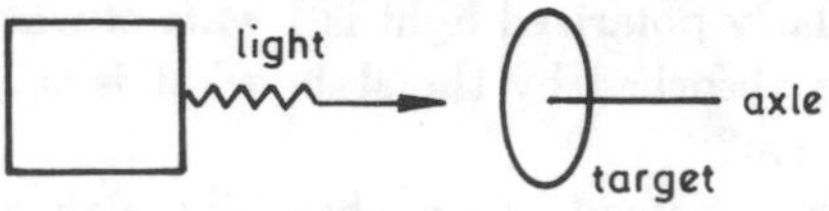

Fig. 2.84

(a) What can you say about the light?

(b) The target is changed from the black surface to a mirror surface. What happens to the torque? Why?

(c) Can you modify the target so as to increase the torque above the values you calculated for cases (a) and (b)? For simplicity keep the target at normal incidence.

(d) Assume visible light (5000 Å) and estimate the torque for the absorbing case (a).

(*UC, Berkeley*)

Solution:

(a) From the fact that the disc absorbs the light totally and starts to spin, we conclude that the light is circularly polarized. A torque

$$\tau_0 = W/\omega$$

is exerted on the disc, where W is the power and ω the angular frequency of the light.

(b) The light is now reflected totally such that the direction of rotation of the plane of polarization of the light in space remains unchanged. No torque is exerted on the target, which remains stationary.

(c) Place a quarter-wave plate before the surface of the mirror in case (b). Passing through the $\lambda/4$ plate twice, the oncoming L (or R)-handed circularly polarized light will be reflected back R (or L)-handed. This will increase the torque to $2\tau_0$. No such enhancement is possible for case (a).

(d) The torque for the absorbing case (a) is

$$\tau = \frac{W}{\omega} = \frac{W\lambda}{2\pi c} = \frac{1 \times 0.5 \times 10^{-6}}{2\pi \times 3 \times 10^8} = 2.65 \times 10^{-16} \text{ N.m}.$$

PART 3 QUANTUM OPTICS

PART 3 QUANTUM OPTICS

3001

A ruby laser emits light with a wavelength of 6943 Å, which to a very good approximation is a plane wave.

(a) Describe qualitatively how such a laser works, including a rough energy level diagram of the relevant atoms involved.

(b) What is the wavelength and frequency of this light as it passes through water (refractive index $n = 4/3$)?

(c) What fraction of each component of polarization of the laser light is reflected as the beam enters the water at 45° to the normal to the surface?

(d) What are the amplitudes of the electric and magnetic field vectors of this plane wave propagating through water, if the time-averaged power of the beam in the water is 100 milliwatts/cm^2?

(e) What is the coherence length of the laser in vacuum (i.e., the distance over which the light stays coherent to 1/4 of a wavelength) if the bandwidth of the laser is $\Delta\nu = 30$ megahertz?

(Columbia)

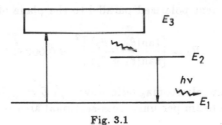

Fig. 3.1

Solution:

(a) The ruby laser is an optically pumped, three-level system. A simplified energy-level diagram is shown in Fig. 3.1. As the ruby is irradiated with a high-intensity flash lamp, the atoms in the ground state E_1 are excited into level E_3, then rapidly relax into the metastable state E_2 by a non-radiative transition because the lifetime at E_3 is very short (about 10^{-9} second). As decay from E_2 is relatively slow, if the energy of the flash lamp is sufficiently intense and optical pumping is applied, a population inversion between E_1 and E_2 occurs. When this condition attains, amplification occurs at the wavelength corresponding to $E_2 - E_1$. Then as the atoms drop down from the level E_2 to the ground level E_1, the characteristic red fluorescent radiation of ruby at this wavelength is emitted as an intense pulse. The optical pumping is accomplished by using two flat mirrors placed at the two ends of the ruby to form a resonant cavity, which reflects to and fro, and amplifies continuously, those spectral modes which propagate along the axis and oscillate with the resonant frequencies of the cavity. These should fall within the frequency bandwidth of the spectral gain of the ruby.

(b) The wavelength in water is

$$\lambda_{water} = \frac{\lambda_{air}}{n} = \frac{6943}{\frac{4}{3}} \approx 5207 \overset{\circ}{A} .$$

(c) The angle of refraction is given by Snell's law to be

$$\theta_2 = \sin^{-1}\left(\frac{3}{4}\sin 45°\right) \approx 32° .$$

The reflectance of the component polarized perpendicular to the plane of incidence is

$$R_\perp = \left[\frac{\sin(\theta_1 - \theta_2)}{\sin(\theta_1 + \theta_2)}\right]^2 \approx 0.05 ,$$

that of the component polarized parallel to the plane of incidence is

$$R_\| = \left[\frac{\tan(\theta_1 - \theta_2)}{\tan(\theta_1 + \theta_2)}\right]^2 \approx 0.0028 .$$

(d) For a plane electromagnetic wave, $\sqrt{\epsilon}E = \sqrt{\mu}H$. The time-averaged power of the beam per unit cross-sectional area is its intensity

$$I = \langle EH \rangle = \sqrt{\frac{\epsilon}{\mu}}\langle E^2 \rangle = \frac{1}{2}\sqrt{\frac{\epsilon}{\mu}}E_0^2 = \frac{n}{2c\mu}E_0^2 ,$$

where E_0 is the amplitude of the electric field, c the velocity of light in free space and n the refractive index of water. The permeability of water is

$$\mu \approx \mu_0 = 4\pi \times 10^{-7} \text{ H/m} .$$

Hence

$$E_0 = \sqrt{\frac{2Ic\mu}{n}} = 7.5 \text{ V/m} ,$$

$$B_0 = \frac{E_0}{v} = \frac{E_0 n}{c} = 3.3 \times 10^{-8} \text{ T} ,$$

B_0 being the amplitude of the magnetic induction.

(e) The uncertainty principle states that

$$\Delta t \cdot \Delta \nu \sim 1,$$

where $\Delta \nu$ is the width of the spectrum line of the laser light, Δt is the time of coherence. Thus the length of coherence is

$$\Delta l = \frac{c}{\Delta \nu} = \frac{3 \times 10^8}{30 \times 10^6} = 10 \text{ m}.$$

3002

A dye laser (Fig. 3.2) consists of two nearly perfectly reflecting mirrors, M, and a gain medium, G, of bandwidth Δf centered at f_0.

(a) What are the allowed frequencies for laser operation in this optical cavity? Express your answer in terms of the time, τ, it takes light to make one round trip in the cavity.

(b) Assume that the laser operates on all possible cavity modes within the gain bandwidth. Also, assume that these modes are all stable in phase, i.e., there are no fluctuations in phase. Also, assume that the phases are adjusted so that all modes are instantaneously in phase at $t = 0$. How will the laser output vary in time?

(c) If it is desired to produce a pulse of one picosecond (10^{-12} sec) duration at a wavelength of 6000 Å, what bandwidth Δf is required? How many laser modes would this involve? (Let $l = 1.5$ m)

(*CUSPEA*)

Fig. 3.2

Solution:

(a) The only modes allowed in the cavity are standing waves for which the wavelengths λ and frequencies f are given by

$$n\lambda = 2l, \quad f = \frac{c}{\lambda} = \frac{cn}{2l} = \frac{n}{\tau},$$

where n is an integer.

(b) Suppose there are N modes within the gain bandwidth. The laser output, under the conditions given, can be expressed as

$$E(t) = \sum_{n}^{N} \left(A_n \cos \frac{2\pi nt}{\tau} + B_n \sin \frac{2\pi nt}{\tau} \right).$$

(c) The last equation shows that the output is a periodic function of time of period τ (see Fig. 3.3). When the phase difference between the highest mode and the lowest mode is 2π, the output ceases. The pulse duration, Δt, is therefore given by

$$\frac{2\pi(N-1)\Delta t}{\tau} = 2\pi, \text{ i.e., } \Delta t \approx \frac{\tau}{N}.$$

The uncertainty principle $\Delta t \cdot \Delta f \sim 1$ gives

$$\Delta f \sim \frac{1}{\Delta t} = 10^{12} \text{ Hz}.$$

As $\lambda f = c$ or $d\lambda = -\frac{\lambda^2}{c} df$, the corresponding band of wavelengths is

$$\Delta \lambda = \Delta f \frac{\lambda^2}{c} \approx 12 \text{ Å}.$$

By definition,

$$\tau = \frac{2l}{c} = 10^{-8} \text{ s},$$

giving

$$N \approx \frac{\tau}{\Delta t} = \frac{10^{-8}}{10^{-12}} = 10^4.$$

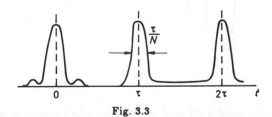

Fig. 3.3

3003

A He-Ne laser operating at 6328 $\overset{\circ}{A}$ has a resonant cavity with plane end mirrors spaced 0.5 m apart. Calculate the frequency separation between the axial modes of this laser. Estimate whether this laser would operate at one or at several axial frequencies given that the linewidth of the Ne 6328 $\overset{\circ}{A}$ line observed in spontaneous emission is typically 0.016 $\overset{\circ}{A}$ wide.

(*Wisconsin*)

Solution:

The axial frequencies of the laser are given by

$$\frac{\lambda n}{2} = l \quad \text{or} \quad f = \frac{nc}{2l},$$

l being the length of the cavity. The frequency separation between successive modes is then

$$\Delta f = \frac{c}{2l}.$$

The frequency bandwidth of the Ne 6328 $\overset{\circ}{A}$ line is given by

$$\Delta f' = c \frac{\Delta \lambda}{\lambda^2},$$

where $\lambda = 6328 \overset{\circ}{A}, \Delta\lambda = 0.016 \overset{\circ}{A}$. Hence

$$\frac{\Delta f'}{\Delta f} = \frac{2l\Delta\lambda}{\lambda^2} = 3.996,$$

the laser will operate at 4 (at least 3) axial frequencies.

3004

The first free-electron laser (FEL) was operated in late 1976 or early 1977.

(a) What does the acronym laser stand for?

(b) What is the output of a laser?

(c) How is the output produced in the FEL? (How is energy fed in? How is it converted to its output form? etc.)

(d) Identify a particular advantage of the FEL over previous types of lasers.

(*Wisconsin*)

Solution:

(a) The acronym laser stands for "Light Amplification by Stimulated Emission of Radiation."

(b) The output of a laser is a highly coherent, both in space and time, and highly directional beam.

(c) As shown in Fig. 3.4, a beam of high energy electrons passes through a transverse, undulating magnetic field; the electrons suffer transverse acceleration and deceleration and electromagnetic radiation is produced by Bremsstrahlung. The initial electromagnetic waves may also be provided by a laser. If a synchronism condition exists between the electron velocity and the phase velocity of the waves, energy may be transferred from the electrons to the waves. The waves are built up in the cavity between two mirrors and finally a laser output is produced.

(d) The advantages of FEL's are the following.

   (i)   FEL is tunable. Varying the energy of the incident electrons allows coherent radiation to be produced at any wavelength from microwaves to visible light, and beyond.

   (ii) It can provide high peak-power, broadband and coherent radiation.

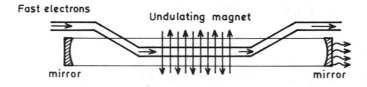

Fig. 3.4

3005

An undulator consists of a linear periodic array of magnets of alternating polarity. Each repeat unit in the array of magnets has length d and there are N such units. An electron of speed v $(v \sim c)$, passing through this undulator, travels a path with only small deflections, as shown in Fig. 3.5. The motion of the electron along this path produces radiation which is most intense at a certain fundamental frequency corresponding to a wavelength λ.

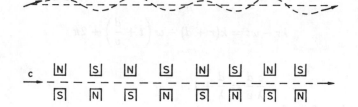

Fig. 3.5

(a) Indicate for which parts of the path electromagnetic radiation in the forward direction will be generated.

(b) Find the wavelength λ using the condition that this forward radiation from successive sections must interfere constructively.

(c) If the undulator consists of N sections of length d, what is the spectral width $\Delta\lambda/\lambda$ of the radiation produced by a beam of mono-energetic electrons?

(d) How does the intensity of the forward radiation produced in the whole undulator compare with that for a single section?

(e) If the velocity of the electrons differs from the speed of light by 1 part in 10^6, calculate the wavelength of the radiation produced by an undulator with $d = 10$ cm.

(*CUSPEA*)

Solution:

(a) As shown in Fig. 3.6, electromagnetic radiation in the forward direction will be generated at points where the tangent to the path is parallel to the axis.

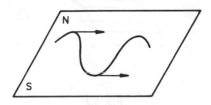

Fig. 3.6

(b) The condition required is

$$kr - \omega t = k(r + d) - \omega\left(t + \frac{d}{v}\right) + 2\pi$$

i.e.,

$$\frac{c}{\lambda}\frac{d}{v} = \frac{d}{\lambda} + 1.$$

Putting $\frac{v}{c} = \beta$, we have

$$\lambda = d\left(\frac{1}{\beta} - 1\right).$$

(c) $\Delta\lambda/\lambda = 1/N$.

(d) $I \approx N^2$.

(e) With $\beta = 1 - 10^{-6}$,

$$\lambda \approx d(\beta - 1) = 10 \times 10^{-6}\text{cm} = 1000\,\overset{\circ}{\text{A}}.$$

3006

A new and important experimental technique being used in several sub-fields of physics is the use of ultra-short light pulses for studying fast phenomena. The ultra-short pulses are produced by mode-locking a laser — a method to be studied in this problem. The model laser consists of an optical cavity of length L with transverse dimensions d formed by two flat mirrors as shown in Fig. 3.7.

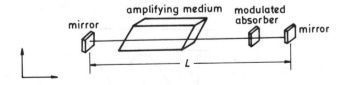

Fig. 3.7

The amplifying medium (it may be solid, liquid, or gas) can be charac-terized by a certain gain spectrum (Fig. 3.8); only light within this band of frequencies will be amplified. To simplify calculation, assume the gain spectrum is rectangular as shown in Fig. 3.9.

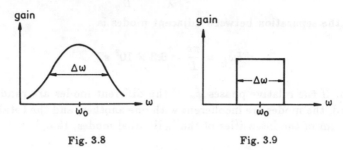

Fig. 3.8   Fig. 3.9

To be specific, consider the amplifying medium Nd glass, which has $\omega_0 = 1.8 \times 10^{15} \, \text{sec}^{-1}$, $\Delta\omega = 6 \times 10^{12} \, \text{sec}^{-1}$, and take the length of the laser to be $L = 150$ cm.

(a) Find the separation $\delta\omega_L$ between adjacent longitudinal modes of the cavity. Neglect the modulator for this part of the problem, neglect dispersion, and, for simplicity, take the refractive index of the amplifying medium to be one.

(b) The rectangular gain spectrum will result in each mode within the gain band having equal amplitude. If the intensity of an individual mode is I_0, what is the total intensity when the relative phases ϕ_n of the different modes are randomly distributed?

(c) By inserting a modulated absorber into the cavity, one can lock the phases of all the modes to the same value, so that traveling waves interfere constructively, producing light pulses. Find the intensity as a function of time in this phase-locked or mode-locked case. What is the pulse duration, pulse separation, and peak intensity (in terms of I_0)? Consider only one polarization.

(It is irrelevant for this problem, but for your information, a modulated absorber has an effective absorption that is a function of time — e.g., $\alpha = \alpha_1 + \alpha_2 \cos(\omega t)$. Any cavity mode with the wrong phase relative to this modulation experiences excess attenuation, so only phase-locked modes are amplified.)

(*Princeton*)

Solution:

(a) The longitudinal modes that can exist in a cavity of length L are those for which

$$L = n\frac{\lambda}{2}, \qquad (n = 1, 2, 3, \ldots),$$

i.e.,

$$\omega_L = \frac{2\pi c}{\lambda} = \frac{\pi n c}{L}.$$

Hence the separation between adjacent modes is

$$\delta\omega_L = \frac{\pi c}{L} = 6.3 \times 10^8 \text{ s}^{-1}.$$

(b) If the relative phases ϕ_n of the different modes are randomly distributed, the modes are incoherent with one another, and the total intensity is the sum of the intensities of the individual modes; that is

$$I = \text{integer}\left(\frac{\Delta\omega}{\delta\omega_L}\right) \cdot I_0 = 9549\, I_0.$$

(c) With mode-locking, the intensity must now be found by first adding up the electric fields, rather than the intensities directly,

$$I(t) = I_0|e^{-i\omega_1 t} + \ldots + e^{-i\omega_n t}|^2 = I_0\left[\frac{\sin\frac{n\delta\omega_L t}{2}}{\sin\frac{\delta\omega_L t}{2}}\right]^2$$

where

$$\omega_n = n\delta\omega_L, \quad n = \text{integer}\left(\frac{\Delta\omega}{\delta\omega_L}\right) = 9549.$$

The pulse duration is

$$\Delta t = \frac{2\pi}{n\delta\omega_L} = \frac{2L}{nc} = 1.05 \times 10^{-12} \text{ s}.$$

The pulse separation is the interval between two maxima:

$$\Delta_t = \frac{2\pi}{\delta\omega_L} = 10^{-8} \text{ s}.$$

The peak intensity occurs at $t = 0$:

$$I(0) = I_0 \cdot n^2 = 9.12 \times 10^7 \, I_0.$$

3007

A projective photon, with frequency f in the rest frame of a target electron, is seen to scatter at $-90°$ with frequency f', while the target scatters at $\theta°$.

(a) Determine the relation of f'/f to θ.

(b) Determine the total electron energy in terms of f, f', and the electron mass m.

(c) If the photon energy loss would be 0.2 of the rest energy of the electron, what would be the speed of the scattered electron?

(d) An observer O moves in a direction parallel to the incident photon's direction at a speed u when the electron-photon collision occurs. What expression would O use for his measure of the electron's energy in its scattered state (in terms of m, v, u and c)?

(SUNY, Buffalo)

Solution:

(a) Referring to Figs. 3.10 and 3.11, we have

$$p_e \sin \theta = p_{f'}, \quad p_e \cos \theta = p_f,$$

where

$$p_e = mv, \quad p_f = \frac{hf}{c}, \quad p_{f'} = \frac{hf'}{c}.$$

Hence $\tan \theta = f'/f$.

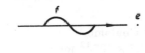

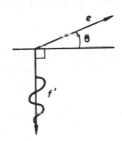

Fig. 3.10   Fig. 3.11

(b) The total energy of the electron is

$$E = [m^2 c^4 + p_e^2 c^2]^{1/2} = [m^2 c^4 + (f^2 + f'^2)h^2]^{1/2}.$$

(c) The rest mass of electron is mc^2 and the photon energy loss is $0.2\, mc^2$. According to the energy conservation law, we have

$$mc^2 + hf = \gamma mc^2 + hf'$$

where

$$\gamma = \frac{1}{\sqrt{1 - \frac{v^2}{c^2}}}.$$

With

$$hf - hf' = (\gamma - 1)mc^2 = 0.2 \; mc^2 \,,$$

we get

$$\gamma = 1.2 \; \text{or} \; v = 0.53 \; c \,.$$

(d) $E' = \dfrac{E - p_0 \cdot u}{\sqrt{1 - u^2/c^2}} = \dfrac{E - p_0 u \cos\theta}{\sqrt{1 - u^2/c^2}} \,,$

where

$$E = \frac{mc^2}{\sqrt{1 - u^2/c^2}} \,, \quad p_0 = hf/c \,.$$

3008

Assume a visible photon of 3 eV energy is absorbed in one of the cones (light sensors) in your eye and stimulates an action potential that produces a 0.07 volt potential on an optic nerve of 10^{-9} F capacitance.

(a) Calculate the charge needed.

(b) Calculate the energy of the action potential.

(Wisconsin)

Solution:

(a) $Q = VC = 0.07 \times 10^{-9} = 7 \times 10^{-11}$ Coulomb.

(b) $E = \frac{QV}{2} = \frac{1}{2} \times 7 \times 10^{-11} \times 0.07 = 2.5 \times 10^{-12}$ joule.

3009

A point source, Q, emits coherent light isotropically at two frequencies, ω and $\omega + \Delta\omega$, with equal power, I joules/sec, at each frequency. Two detectors, A and B, each with a (small) sensitive area S capable of responding to individual photons, are located at distances l_A and l_B from Q as shown in Fig. 3.12. In the following, take $\frac{\Delta\omega}{\omega} \ll 1$ and assume the experiment is carried out in vacuum.

(a) Calculate the individual photon counting rates (potons/sec) at A and at B as functions of time. Consider time scales $\gg 1/\omega$.

(b) If now the output pulses from A to B are put into a coincidence circuit of resolving time τ, what is the time-averaged coincidence counting rate? Assume that $\tau \ll 1/\Delta\omega$ and recall that a coincidence circuit will produce an output pulse if the two input pulses arrive within a time τ of one another.

(CUSPEA)

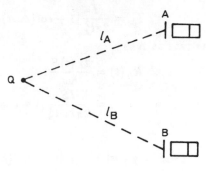

Fig. 3.12

Solution:

(a) The waves of angular frequencies ω_1 and ω_2 arriving at A can be represented by

$$E_1 = \frac{E_0}{l_A} e^{i(\omega_1 t - k_1 l_A)}$$

$$E_2 = \frac{E_0}{l_A} e^{i(\omega_2 t - k_2 l_A)}$$

respectively, where $\omega_1 = \omega, \omega_2 = \omega + \Delta\omega$. The resultant wave at A is then

$$E_A = E_1 + E_2 = (E_0/l_A)[e^{i(\omega_1 t - k_1 l_A)} + e^{i(\omega_2 t - k_2 l_A)}]$$

and the intensity at A, time-averaged, is

$$I_A \sim \frac{1}{2} E_A E_A^* = \left(\frac{E_0}{l_A}\right)^2 \{1 + \cos[(\omega_2 - \omega_1)t - (k_2 - k_1)l_A]\}.$$

The constant phase difference $(k_2 - k_1)l_A$ can be taken as zero by choosing a suitable time origin. The intensity of the ω_1 wave at A is

$$I_1 \sim \frac{1}{2} E_1 E_1^* = \frac{1}{2}\left(\frac{E_0}{l_A}\right)^2.$$

As the source has power I and is isotropic,

$$I_1 = \frac{I}{4\pi l_A^2},$$

giving

$$\left(\frac{E_0}{l_A}\right)^2 \sim \frac{I}{2\pi l_A^2}.$$

Hence

$$I_A = \frac{I}{2\pi l_A^2}[1 + \cos(\Delta\omega t)].$$

The photon counting rate at A is

$$R_A(t) = \frac{I_A(t)S}{\hbar\omega}.$$

$$= \frac{IS}{2\pi\hbar\omega l_A^2}[1 + \cos(\Delta\omega t)].$$

Similarly, we have

$$I_B(t) = \frac{I}{2\pi l_B^2}\{1 + \cos[\Delta\omega t + (l_B - l_A)(k_1 - k_2)]\}.$$

As

$$k_1 = \frac{\omega_1}{c} = \frac{\omega}{c},$$

$$k_2 = \frac{\omega_2}{c} = \frac{\omega + \Delta\omega}{c},$$

$$k_1 - k_2 = -\frac{\Delta\omega}{c},$$

we have

$$I_B(t) = \frac{I}{2\pi l_B^2}\left\{1 + \cos\left[\Delta\omega t + \frac{(l_A - l_B)}{c}\Delta\omega\right]\right\},$$

giving

$$R_B(t) = \frac{IS}{2\pi\hbar\omega l_A^2}\left\{1 + \cos\left[\Delta\omega t + \frac{(l_A - l_B)\Delta\omega}{c}\right]\right\}.$$

(b) The probability that a signal from A at time t be accompanied by one from B within a time $\pm\tau$ is $2\tau R_B(t)$. Hence the coincidence counting rate is

$$R_{AB}(t) = R_A(t)R_B(t)(2\tau)$$

$$= \frac{I^2 S^2 \tau}{2\pi^2\hbar^2\omega^2 l_A^2 l_B^2}\left[1 + \cos(\Delta\omega t)\right]\left\{1 + \cos\left[\Delta\omega\left(t + \frac{l_A - l_B}{c}\right)\right]\right\}.$$

3010

Describe briefly how a photomultiplier tube works. Can such a tube be used to distinguish between two photons whose energies differ by 50%?

(Columbia)

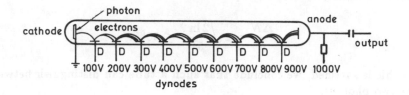

Fig. 3.13

Solution:

A photomultiplier tube is a device which multiplies a single photo-electron by five or six orders of magnitude. In addition to a photo-emission cathode, it has several dynodes which are maintained at successively higher potentials (see Fig. 3.13). Generally the potential is increased by 100-volt steps from one dynode to the next. The potential difference accelerates the electrons so that when they strike the surface of a dynode, each causes three or four secondary electrons to be ejected. These secondary electrons are then accelerated towards the next dynode and the process is repeated. In this manner, a single electron at the cathode can be multiplied to one million electrons at the anode after ten stages.

The resultant anode current I, which is proportional to the mean number $\overline{N}$ of the electrons collected, is on average proportional to the energy E of the incoming photon. The number N of the electrons collected satisfies the Poisson distribution

$$P(N) = \overline{N}^N e^{-N}/N!.$$

For large $\overline{N}$, the Poisson distribution approximates the Gaussian distribution:

$$P(N) \approx \frac{1}{\sqrt{2\pi}\sigma} \exp\left[-\frac{(N-\overline{N})^2}{2\sigma^2}\right], \ \sigma = \sqrt{\overline{N}}.$$

The half-width of the distribution, ΔN, (see Fig. 3.14) is given by

$$P\left(\overline{N} + \frac{\Delta N}{2}\right) \bigg/ P(\overline{N}) = \frac{1}{2} = \exp\left[-\frac{(\Delta N/2)^2}{2\sigma^2}\right],$$

i.e.,

$$\Delta N = 2\sqrt{2\ln 2 \cdot \sigma^2} = 2\sqrt{2\overline{N}\ln 2}.$$

For a tube with 10 dynodes, each of secondary-emission rate 5, $\overline{N} \simeq 5^{10}$. From Fig. 3.15, we see that to distinguish two photons of energies E_1

and E_2 differing by 50% we require

$$\frac{\Delta N}{\overline{N}} \approx 2\sqrt{\frac{2\ln 2}{5^{10}}} \leq 50\%.$$

As this is satisfied, we conclude that such a tube can distinguish between the two photons.

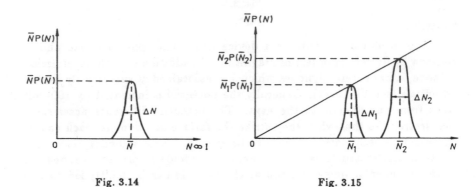

Fig. 3.14   Fig. 3.15

3011

(a) What is meant by
    (i) the Doppler line width, and
    (ii) the natural line width of a spectral line?

(b) Describe an experiment for making "Doppler free" measurements of spectral lines. (You need not be quantitative, but your answer should make clear that you understand the basic physics involved.)

(Wisconsin)

Solution:

(a) (i) The Doppler broadening of spectral lines is caused by the random thermal motion of the radiating atoms and is proportional to $T^{1/2}$, where T is the absolute temperature of the source.

(ii) According to Heisenberg's uncertainty principal, $\Delta t \cdot \Delta E \sim h$, an excited state of lifetime Δt of an atom has an uncertainty in energy of ΔE. The corresponding spectral line $\nu = \frac{E}{h}$ therefore has a natural width $\Delta \nu = \frac{\Delta E}{h} \sim \frac{1}{\Delta t}$.

(b) The method of double-photon absorption can be employed to make "Doppler free" measurements of spectral lines. A molecule, having absorbed two photons $\hbar\omega_1(\mathbf{k}_1)$ and $\hbar\omega_2(\mathbf{k}_2)$, would be exited from state i to state f. If the molecule is motionless,

$$E_f - E_i = \hbar(\omega_1 + \omega_2),$$

and no Doppler shift occurs. If the molecule moves with velocity $\mathbf{v}$,

$$E_f - E_i = \hbar[\omega_1 + \omega_2 - \mathbf{v} \cdot (\mathbf{k}_1 + \mathbf{k}_2)],$$

and a Doppler shift $\mathbf{v} \cdot (\mathbf{k}_1 + \mathbf{k}_2)$ occurs. The Doppler shift is zero if $\mathbf{k}_1 = -\mathbf{k}_2$.

Two beams of laser light of the same frequency ω are incident in opposite directions on a gas sample as shown in Fig. 3.16. Double-photon absorption that takes place can be detected by the $f \to m$ (an intermediate state) fluorescent transition.

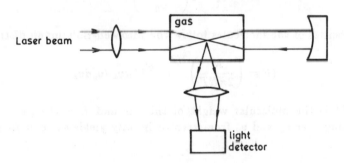

Fig. 3.16

3012

(a) Consider the emission or absorption of visible light by the molecules of a hot gas. Derive an expression for the frequency distribution $f(\nu)$ expected for a spectral line, central frequency ν_0, due to the Doppler broadening. Assume an ideal gas at temperature T with molecular weight M.

Consider a vessel filled with argon gas at a pressure of 10 Torr(1 Torr = 1 mm of mercury) and a temperature of 200 °C. Inside the vessel is a small piece of sodium which is heated so that the vessel will contain some sodium vapor. We observe the sodium absorption line at 5896 Å in the light from a tungsten filament passing through the vessel. Estimate:

(b) The magnitude of the Doppler broadening of the line.

(c) The magnitude of the collision broadening of the line.

Assume here that the number of sodium atoms is very small compared to the number of argon atoms. Make reasonable estimates of quantities that you may need which are not given and express your answers for the broadening in Angstroms.

Atomic weight of sodium = 23.   (CUSPEA)

Solution:

(a) The Doppler frequency shift, viewed along the z direction, is given by

$$\nu = \nu_0\left(1 + \frac{v_z}{c}\right),$$

i.e.,

$$v_z = c\left(\frac{\nu - \nu_0}{\nu_0}\right).$$

The velocities of the molecules follow the Maxwell-Boltzmann distribution

$$dP = \left(\frac{M}{2\pi RT}\right)^{3/2} e^{-\frac{Mv^2}{2RT}}\,dv_x dv_y dv_z,$$

where M is the molecular weight of the gas and R is the gas constant. Integrating over v_x and v_y from zero to infinity yields the distribution for v_z:

$$dP = \left(\frac{M}{2\pi RT}\right)^{1/2} e^{-\frac{Mv_z^2}{2RT}}\,dv_z,$$

giving the frequency distribution

$$dP = \frac{1}{\nu_0}\left(\frac{Mc^2}{2\pi RT}\right)^{1/2} e^{-Mc^2(\nu-\nu_0)^2/2RT\nu_0^2}\,d\nu.$$

(b) The frequency distribution has the form of a Gaussian distribution with the standard deviation σ give by

$$\sigma^2 = \frac{RT\nu_0^2}{Mc^2}.$$

Hence

$$\Delta\nu \approx \sqrt{\frac{RT}{Mc^2}}\,\nu_0,$$

or

$$\Delta\lambda \approx \sqrt{\frac{RT}{Mc^2}}\lambda_0 = \sqrt{\frac{(8.3)(473)}{(40 \times 10^{-3})}\frac{\lambda_0}{(3 \times 10^8)}}$$

$$= 1.04 \times 10^{-6}\lambda_0 = 1.04 \times 10^{-6} \times 5896\,\overset{\circ}{A} = 6.13 \times 10^{-3}\,\overset{\circ}{A}.$$

(c) The mean free path Λ is given by

$$\Lambda = \frac{1}{n\sigma},$$

where $\sigma = \pi r^2 \approx 3 \times 10^{-20}$ m^2, taking $r \approx 10^{-10}$ m. The number of molecules per unit volume is

$$n = \frac{A}{V} = \frac{pA}{RT},$$

where A is Avogadro's number. Thus

$$n = \frac{(1.01 \times 10^5 \times 10/760)}{(8.3)(473)} \times 6.02 \times 10^{23}$$

$$= 2.04 \times 10^{23} \text{ m}^{-3}.$$

Substituting gives $\Lambda = 1.7 \times 10^{-4}$ m. The average time interval between two collisions is $\tau = \Lambda/v$, where the v is the mean speed of a sodium atom,

$$v \approx \sqrt{\frac{RT}{M}} = \sqrt{\frac{(8.3)(473)}{23 \times 10^{-3}}} \approx 413 \text{ ms}^{-1}.$$

Thus

$$\tau = \frac{1.7 \times 10^{-4}}{413} = 4 \times 10^{-7}\text{s}.$$

The broadening due to collision is given by the collision frequency $\Delta\nu = 1/\tau$. In terms of wavelength, the broadening is

$$\Delta\lambda = \frac{c}{\nu^2}\Delta\nu = \frac{\lambda^2}{c\tau} = 3 \times 10^{-5}\,\overset{\circ}{A}.$$

3013

Consider a gas of atoms A at $T = 300$ K, $p = 100$ Torr. The mass of each atom is 4.2×10^{-27} kg and Boltzmann's constant is 1.4×10^{-23} J/K. Some of the atoms are in excited states A^* and emit radiation of frequency ν.

(a) Estimate the Doppler width $\Delta\nu_D/\nu$.

(b) Assume a reasonable cross-section for A^*-A collision and estimate the pressure-broadened width of the line $\Delta\nu_p/\nu$.

(*Wisconsin*)

Solution:

(a)

$$\frac{\Delta\nu_D}{\nu} = \frac{1}{c}\sqrt{\frac{kT}{m}} = \frac{1}{3 \times 10^8}\sqrt{\frac{1.4 \times 10^{-23} \times 300}{4.2 \times 10^{-27}}} = 3 \times 10^{-6}.$$

(b) The pressure-broadened width $\Delta\nu_p$ is the collision frequency of a molecule of the gas

$$\Delta\nu_p = v/\Lambda,$$

where v is its mean speed, given by $\sqrt{kT/m}$, and Λ is the mean free path of a molecule, given by

$$\frac{1}{\pi r^2 n} = \frac{1}{\pi r^2}\sqrt{\frac{kT}{p}},$$

r being the radius of a molecule $\approx 10^{-10}$ m, n the number of molecules per unit volume. Thus

$$\Delta\nu_p = \pi r^2 n\sqrt{\frac{kT}{m}} = \frac{\pi p r^2}{\sqrt{kmT}} = 7.2 \times 10^7 \text{ Hz}.$$

For visible light, $\lambda \approx 5000$ Å or $\nu = 6 \times 10^{14}$ Hz,

$$\Delta\nu_p/\nu = 1.2 \times 10^{-7}.$$

3014

An electronic transition in ions of ^{12}C leads to photon emission near $\lambda = 500$ nm ($h\nu = 2.5$ eV). The ions are in thermal equilibrium at a temperature $kT = 20$ eV, a density $n = 10^{24}/\text{m}^3$, and a non-uniform magnetic field which ranges up to $B = 1$ Tesla.

(a) Briefly discuss broadening mechanisms which might cause the transition to have an observed width $\Delta\lambda$ greater than that obtained for very small values of T, n and B.

(b) For one of these mechanisms calculate the broadened width $\Delta\lambda$ using order of magnitude estimates of the needed parameters.

(*Wisconsin*)

Solution:

(a) Thermal motion of ions broadens spectral lines because of the Doppler effect. The broadening is proportional to $\sqrt{T}$, where T is the absolute temperature.

Collisions between ions also cause a broadening, the collision or pressure broadening, which is proportional to $\sqrt{T}$ and to n, the number density of ions, or p, the pressure of the gas.

When a source is placed in a magnetic field, a spectral line may break up with the energy shifts proportional to B, the strength of the magnetic field. This spliting is known as the Zeeman effect. If the magnetic field is non-uniform, the spectral shifts would range from zero to a maximum value dependent on B, i.e., the spectral lines are broadened with the width proportional to B.

(b) The Doppler broadening is given by

$$\frac{\Delta\lambda}{\lambda} = \frac{1}{c}\sqrt{\frac{kT}{m}} = \frac{1}{3\times10^8}\left(\frac{20\times1.6\times10^{-19}}{12\times1.67\times10^{-27}}\right)^{1/2}$$

$$= \frac{1.3\times10^4}{3\times10^8} \approx 4\times10^{-5}.$$

Thus

$$\Delta\lambda = 5000\times4\times10^{-5} = 0.2\,\overset{\circ}{A}.$$

The collision broadening is given by

$$\Delta\nu \approx r^2 n\sqrt{\frac{kT}{m}} = 10^{-20}\times10^{24}\times1.3\times10^4 = 1.3\times10^8 \text{ s}^{-1},$$

or

$$\Delta\lambda = \lambda^2\Delta\nu/c \approx 0.1\,\overset{\circ}{A}.$$

The energy shift of a spectral line caused by the Zeeman effect is given by

$$\Delta E = \mu_B B\Delta(Mg),$$

where M is the magnetic quantum number, g the Landé factor, μ_B the Bohr magneton

$$\mu_B = \frac{eh}{4\pi mc} = 0.93 \times 10^{-23} \text{ A} \cdot \text{m}^{-2}.$$

Taking $\Delta(Mg) \sim 1$, we have, with $B = 1$ Tesla,

$$\Delta E \approx 10^{-23} \text{ J}.$$

Hence

$$\Delta\lambda = \frac{\lambda^2 \Delta\nu}{c} = \frac{\lambda^2 \Delta E}{hc} \approx \left(\frac{5000^2 \times 10^{-23}}{6 \times 10^{-34} \times 3 \times 10^{18}}\right) = 0.1 \overset{\circ}{\text{A}}.$$

3015

During the days when there are no clouds and the atmosphere is relatively dust (smog) free and has low humidity, the sky, away from the sun, has a very deep blue color. This is due to molecular scattering.

(a) If the molecular electric polarizability due to the photon E field is frequency independent, show how this leads to a wavelength dependence of the scattering of sunlight similar to that observed. Discuss the angular dependence of the scattering.

(b) If in each air volume element ΔV of sides small compared with λ, the mean number of molecules is $N_0 \Delta V \gg 1$, and if there is no fluctuation about this mean number, show that no net scattering results.

(c) If the root-mean-square fluctuation of the number of molecules equals the square root of the mean number, show that the net scattering intensity is as if each molecule scattered independently with no interference between molecules.

(d) When the relative humidity is near 100% there is observed to be enhanced scattering (relative to (c)) for water molecules. Explain.

(*Columbia*)

Solution:

(a) The electric field $\mathbf{E} = \mathbf{E}_0 \cos(\omega t)$ of the incident light polarizes the air molecules into oscillating dipoles, each of moment $\mathbf{p}_0 \cos(\omega t)$, where

$$\mathbf{p}_0 = \frac{e^2}{m} \frac{\mathbf{E}_0}{(\omega_0^2 - \omega^2)},$$

e and m being the charge and mass of an electron, ω_0 its characteristic frequency. For optical frequencies, $\omega \ll \omega_0$, so that the polarizability $e^2/m\omega_0^2$ is independent of frequency.

The field of the radiation scattered by such an induced dipole has an amplitude at a large distance r from the dipole of

$$E = \frac{1}{4\pi\epsilon_0}\left(\frac{\omega}{c}\right)^2 \frac{\sin\theta}{r}p_0,\tag{1}$$

where θ is the angle between $\mathbf{r}$ and the orientation of the dipole. The corresponding scattered intensity is

$$
\begin{aligned}
I &= \frac{1}{2}EH = \frac{1}{2}\sqrt{\frac{\epsilon_0}{\mu_0}}E^2 = \frac{\omega^4}{32\pi^2\epsilon_0 c^3}\left(\frac{\sin\theta}{r}\right)^2 p_0^2 \\
&= \left(\frac{e^2}{4\pi\epsilon_0 mc^2}\right)^2 \frac{\epsilon_0 c E_0^2}{2}\left(\frac{\omega}{\omega_0}\right)^4\left(\frac{\sin\theta}{r}\right)^2 \\
&= a^2 I_0 \left(\frac{\omega}{\omega_0}\right)^4\left(\frac{\sin\theta}{r}\right)^2,
\end{aligned}\tag{2}
$$

where a is the classical radius of an electron and I_0 the intensity of the incident light. Thus $I \propto \omega^4 \propto \frac{1}{\lambda^4}$. Hence, of the visible light the blue color is preferentially scattered. This accounts for the blueness of the sky if we are not watching along the line of sight of the sun.

The scattered intensity has an angular distribution $\sim \sin^2\theta$. For an unpolarized incident light, let its direction be along the z-axis and consider light scattered into a direction in the xz plane making angle ψ with the z-axis as shown in Fig. 3.17(a). At some instant, the electric field of the incident light at the dipole, $\mathbf{E}_0$, (and hence the induced dipole) makes angle ϕ with the x-axis. Geometry gives

$$\cos\theta = \cos\phi\sin\psi.$$

As ϕ is random, we ahave to average $\sin^2\theta$ over ϕ:

$$\langle\sin^2\theta\rangle = \frac{1}{2\pi}\int_0^{2\pi}(1 - \cos^2\phi\sin^2\psi)d\phi = \frac{1}{2}(1 + \cos^2\psi).$$

Hence the scattered intensity for an unpolarized incident light has an angular distribution $\sim (1 + \cos^2\psi)$. The scattered intensity is maximum in the forward and backward directions, and is minimum in a transverse direction.

(b) Consider a small volume element $\Delta V \ll \lambda^3 \ll r^3$. Imagine it as being made up of q equal layers, each of thickness $\ll \lambda$, $\sqrt{2}\lambda/q$ apart and making an angle $\frac{\pi}{4}$ with the incident light as shown in Fig. 3.17(b). The phase difference between any two adjacent layers is

$$\Delta\delta = \frac{2\pi}{\lambda} \frac{\sqrt{2}\lambda}{q} \sin\frac{\pi}{4} = \frac{2\pi}{q}.$$

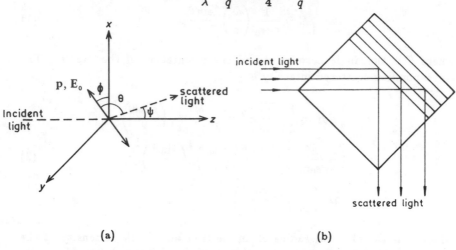

(a)   (b)

Fig. 3.17

Let the field of the light scattered in the direction ψ by the k-th layer be $E_k \cos(\omega t - \delta_k)$. The resultant field in that direction is then

$$E = \sum_{k=1}^{q} E_k \cos(\omega t - \delta_k)$$

with

$$\delta_k - \delta_{k-1} = \frac{2\pi}{q}.$$

If the molecules are uniformly distributed in the volume element with no fluctuation, then $E_1 = E_2 = \ldots = E_q$. Hence

$$E = E_1 \sum_{k=0}^{q-1} \cos\left(\omega t - \delta_1 + \frac{2k\pi}{q}\right)$$

$$= E_1 \mathrm{Re}\left\{ e^{i(\omega t - \delta_1)} \sum_{k=0}^{q-1} e^{i\frac{2k\pi}{q}} \right\}$$

$$= E_1 \mathrm{Re}\left\{ e^{i(\omega t - \delta_1)} \left(\frac{1 - e^{i2\pi}}{1 - e^{i\frac{2\pi}{q}}} \right) \right\} = 0.$$

Therefore, no net scattering results.

(c) If the number N of molecules in each layer fluctuates about a mean $\overline{N}$, the E_k's also fluctuate about a mean $\overline{E}$:

$$E_k = \overline{E} + \Delta_k \,.$$

Then

$$E = \sum_{k=1}^{q} (\overline{E} + \Delta_k) \cos(\omega t - \delta_k)$$

$$= \sum_{k=1}^{q} \Delta_k \cos(\omega t - \delta_k)$$

so that

$$E^2 = \sum_{k=1}^{q} \Delta_k^2 \cos^2(\omega t - \delta_k)$$

$$+ \sum_{k=1}^{q} \sum_{l \neq k} \Delta_k \Delta_l \cos(\omega t - \delta_k) \cos(\omega t - \delta_l)$$

$$\approx \sum_{k=1}^{q} \Delta_k^2 \cos^2(\omega t - \delta_k)$$

since the Δ_k's are independent of one another and are as likely positive as negative.

We can write

$$E^2 \approx q \langle \Delta_k^2 \rangle \cos^2(\omega t - \delta) \,.$$

From Eq. (1) we see that

$$\Delta_k^2 = \left(\frac{1}{4\pi\epsilon_0} \right)^2 \left(\frac{\omega}{c} \right)^4 \left(\frac{\sin\theta}{r} \right)^2 (N_k - \overline{N})^2 p_0^2 \,.$$

Then

$$\langle \Delta_k^2 \rangle = \left(\frac{1}{4\pi\epsilon_0} \right)^2 \left(\frac{\omega}{c} \right)^4 \left(\frac{\sin\theta}{r} \right)^2 \overline{N} p_0^2 \,,$$

as it is given that $\langle (N_k - \overline{N})^2 \rangle = \overline{N}$.

Hence

$$I = \frac{1}{2} \sqrt{\frac{\epsilon_0}{\mu_0}} q \langle \Delta_k^2 \rangle$$

$$= \frac{\omega^4}{32\pi^2 \epsilon_0 c^3} \left(\frac{\sin\theta}{r} \right)^2 q \overline{N} p_0^2 \,.$$

Now $q\overline{N}$ is the total number of molecules in ΔV. A comparison of the above with Eq. (2) shows that the net scattered intensity is the same as if each of the molecules scattered independently with no interference between them.

For a dilute gas of volume V and N_0 moleclules per unit volume, we then have from Eq. (2)

$$I = \left(\frac{e^2}{4\pi\epsilon_0 mc^2}\right) I_0 \left(\frac{\omega}{\omega_0}\right)^4 \left(\frac{\sin\theta}{r}\right)^2 N_0 V \ .$$

If its refractive index is n, the Clausius-Mossotti relation

$$n^2 - 1 = \frac{N_0 e^2}{\epsilon_0 m\omega_0^2}$$

gives

$$I = \pi^2 I_0 \frac{(n^2 - 1)^2}{\lambda^4} \frac{V}{N} \left(\frac{\sin\theta}{r}\right)^2 ,$$

which is known as Rayleigh's formula.

(d) When the relative humidity is near 100%, water molecules condense to form droplets. These fine particles scatter the blue color when a beam of white light passes through them. This is known as Tyndall scattering and is much stronger than Rayleigh scattering.

3016

On a clear day the sky appears blue because of:

(a) reflection from the sea,

(b) density fluctuations of the atmosphere which cause scattering,

(c) cobalt vapors in the atmosphere.

(*CCT*)

Solution:

The answer is (b).

3017

A typical molecular gas shows absorption bands over the entire electromagnetic spectrum from X-rays to radio waves. In terms of atomic and molecular structure state the absorption process responsible for absorption bands in the following regions of the spectrum:

(a) X-ray,

(b) ultraviolet and visible,

(c) near infra-red,

(d) far infra-red and radio.

(*Wisconsin*)

Solution:

When electromagnetic waves pass through a gas into a spectroscope, those wavelengths which the medium would emit if its temperature were raised high enough will be absorbed, so that the spectrum will contain dark lines on regions corresponding to upward transitions between the energy levels of the atoms or molecules of the gas.

(a) X-ray: Removal of an electron from the inner shell of an atom to an outer shell or out of the atom.

(b) Ultraviolet and visible: Transition between the energy states of an orbiting atomic electron.

(c) Near infra-red: Transition between the vibrational levels of a molecule.

(d) Far infra-red: Transition between the rotational levels of a molecule.

3018

Consider the earth's atmosphere (consisting of $O_2, N_2, CO_2, N_2O, H_2O$ O_3 etc.). Discuss whether the atmosphere is reasonably transparent or strongly absorbing for each of the following frequency regions. If absorbing, say what are the most important mechanisms.

(a) $10^8 - 10^9$ Hz.

(b) far infra-red

(c) near infra-red

(d) visible

(e) ultra-violet

(f) X-rays

(g) γ-rays

(*Columbia*)

Solution:

(a) Microwaves of frequencies $10^8 - 10^9$ Hz are strongly absorbed by O_2, H_2O and N_2O.

(b) Far infra-red light is strongly absorbed by CO_2.

(c) Near infra-red light is strongly absorbed by H_2O.

(e) Ultra-violet light is strongly absorbed by O_3.

(d), (f), (g) The atmosphere is transparent to X-rays, γ-rays and the visible light, though some spectral lines of the last are absorbed by O_3.

3019

Explain the physical principles involved in the following phenomena:

(a) the blue sky

(b) the red sun at sunset

(c) the rainbow

(e) the twinkling of stars.

(Columbia)

Solution:

(a) The intensity of the radiation scattered by air molecules is, according to Rayleigh's scattering formula, inversely proportional to the fourth power of the wavelength. Hence, the shorter wavelength (blue color) component of the visible sun light is more strongly scattered and the sky appears blue when the observation is not along the line of illumination (see Fig. 3.18(a)).

(b) At sunset or sunrise, the sun light passes through a thicker layer of the atmosphere than at noon and its blue components are scattered out. Consequently the sun looks reddish (see Fig. 3.18(b)).

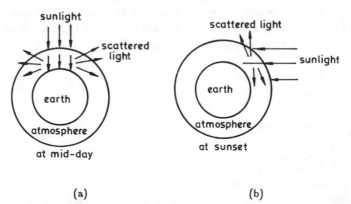

(a)   (b)

Fig. 3.18

(c) A quantity of small water droplets remains in the atmosphere after rain which refract the sunlight. As the index of refraction of water is dependent on wavelength, components of sunlight of different wavelengths or colors are refracted into different angles, giving rise to the rainbow (see Fig. 3.19).

(d) The random fluctuation of the density of air causes the twinkling of stars.

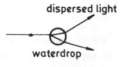

Fig. 3.19

3020

Sunlight travels to earth in approximately:
- (a) 1 hr
- (b) 8 min
- (c) 8 sec

(*CCT*)

Solution:

The answer is (b).

3021

Sunlight is normally incident on the surface of water with index of refraction $n = 1.33$.

(a) Derive the energy reflection and transmission coefficients, R and T. $(R + T = 1)$

(b) If the incident flux is 1 kW/m^2, what is the pressure that sunlight exerts on the surface of the water?

Be careful: part (b) may not follow from part (a) as directly as you think!

(*UC, Berkeley*)

Solution:

(a) Fresnel's formulae give for normally incidence

$$R = \left(\frac{n_2 - n_1}{n_2 + n_1}\right)^2 = \left(\frac{n - 1}{n + 1}\right)^2 = 0.02,$$

$$T = \frac{4n_1 n_2}{(n_2 + n_1)^2} = \frac{4n}{(n + 1)^2} = 0.98,$$

$$R + T = 1.$$

(b) From the point of view of the photon theory, light pressure is the result of a transfer of the momentum of the incoming light to the target. If W is the incident flux density and R the reflection coefficient, the change of momentum per unit area is

$$\frac{WR}{c} - \left(-\frac{W}{c}\right) = \frac{(1 + R)W}{c}.$$

By definition this is the light pressure p

$$p = \frac{(1 + 0.02) \times 1000}{3 \times 10^8} = 3.34 \times 10^{-6} \text{ N/m}^2.$$

3022

An electromagnetic wave $(\mathbf{E}_0, \mathbf{H}_0)$ traveling in a medium with index of refraction n_1 strikes a planar interface with a medium with n_2. The angle the wave makes with the normal to the interface is θ_1. Derive the general Fresnel equations for the reflected and refracted waves. Obtain Brewster's law as a special case of these equations.

(Columbia)

Solution:

If the incident electromagnetic wave is unpolarized, it can be considered as consisting of two components, one plane polarized with the electric field in the plane of incidence, and the other plane polarized with the electric field perpendicular to the plane of incidence. Consider the first component and let the electric and magnetic vectors of the incident, reflected and refracted waves be represented by $E_{p_1}, E'_{p_1}, E_{p_2}$ and $H_{s_1}, H'_{s_1}, H_{s_2}$, respectively (See Fig. 3.20). The boundary conditions on the field vectors give

$$E_{p_1} \cos\theta_1 + E'_{p_1} \cos\theta_1 = E_{p_2} \cos\theta_2 \,, \tag{1}$$

$$H_{s_1} - H'_{s_1} = H_{s_2} \,. \tag{2}$$

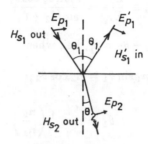

Fig. 2.20

For a plane electromagnetic wave,

$$\sqrt{\epsilon_1} E_{p_1} = \sqrt{\mu_1} H_{s_1} \,,$$
$$\sqrt{\epsilon_1} E'_{p_1} = \sqrt{\mu_1} H'_{s_1} \,,$$
$$\sqrt{\epsilon_2} E_{p_2} = \sqrt{\mu_2} H_{s_2} \,.$$

Assuming the media to be non-ferromagnetic, we have

$$\mu_1 = \mu_2 = \mu_0 \,,$$

and so

$$n_1 = \sqrt{\frac{\epsilon_1}{\epsilon_0}} \,, \quad n_2 = \sqrt{\frac{\epsilon_2}{\epsilon_0}} \,.$$

Equation (2) can then be written as

$$n_1 E_{p_1} - n_1 E'_{p_1} = n_2 E_{p_2} \,. \tag{3}$$

Snell's law gives $n_1 \sin\theta_1 = n_2 \sin\theta_2$. Combining this with Eqs. (1) and (2) yields

$$r_p \equiv \frac{E'_{p_1}}{E_{p_1}} = \frac{\tan(\theta_2 - \theta_1)}{\tan(\theta_2 + \theta_1)} \,, \quad t_p \equiv \frac{E_{p_2}}{E_{p_1}} = \frac{2 \sin\theta_2 \cos\theta_1}{\sin(\theta_2 + \theta_1) \cos(\theta_2 - \theta_1)} \,,$$

where r_p and t_p are called the amplitude reflection and transmission coefficients respectively.

Similarly, for the incident component with its electric field vector perpendicular to the plane of incidence we have

$$r_s \equiv \frac{E'_{s_1}}{E_{s_1}} = \frac{\sin(\theta_2 - \theta_1)}{\sin(\theta_2 + \theta_1)} \,,$$

$$t_s \equiv \frac{E_{s_2}}{E_{s_1}} = \frac{2\sin\theta_2\cos\theta_1}{\sin(\theta_2 + \theta_1)} \,.$$

Note that $r_p = 0$ when $\theta_1 + \theta_2 = \frac{\pi}{2}$, i.e., when

$$n_1 \sin\theta_1 = n_2 \cos\theta_2 \,,$$

or

$$\theta_1 = \tan^{-1}\left(\frac{n_2}{n_1}\right) \,.$$

This is stated as Brewster's law: When the incident angle is equal to $\tan^{-1}(\theta_2/\theta_1)$, the component with **E** in the plane of incidence is not reflected and an unpolarized wave becomes plane polarized on reflection.

3023

Calculate the ratio of the intensities of the reflected wave and the incident wave for a light wave incident normally on the surface of a deep body of water with index of refraction $n = 1.33$.

(*Wisconsin*)

Solution:

Fresnel's equations give for light incident normally

$$R = \frac{I_r}{I_i} = \left(\frac{n_2 - n_1}{n_2 + n_1}\right)^2 = 0.02 \,.$$

3024

While an aquarium is being filled with water (index of refraction $n > 1$), a motionless fish looks up vertically through the rising surface of the water at a stationary monochromatic plane wave light source. Does the fish see the light source blue-shifted (higher frequency), red-shifted (lower frequency), or is the frequency change identically zero? Explain your reasoning.

Solution:

The speed of light propagating through a moving medium of index of refraction n with speed V relative to an observer was given by Fresnel as

$$u = \frac{c}{n} - V\left(1 - \frac{1}{n^2}\right),$$

which has been experimentally verified by Fizeau. The wavelength of light propagating through water is

$$\lambda_{\text{water}} = \frac{c}{n\nu},$$

where c and ν are the speed and frequency of light in vacuum. The frequency of light as seen by the fish is

$$\nu' = \frac{u}{\lambda_{\text{water}}} = \nu - V\left(1 - \frac{1}{n^2}\right)\left(\frac{n\nu}{c}\right).$$

For $n > 1, \nu' < \nu$, the fish would see the light red-shifted.

3025

Solar energy at the rate 800 W/m^2 strikes a flat solar panel for water heating. If the panel has absorptance = 0.96 for all wavelengths and the sides are perfect insulators, calculate the maximum temperature of the water. If the absorptance dropped by 1/2, how would this affect the final temperature?

(*Wisconsin*)

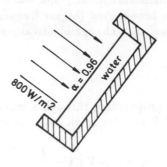

Fig. 3.21

Solution:

The Stefan-Boltzmann law gives, under equilibrium,

$$\alpha\phi = \sigma T^4$$

where $\sigma = 5.67 \times 10^{-8}$ W/m^{2}K^4 is Stefan's constant, ϕ is the solar energy flux density, and α is the absorptance of the panel. With

$$\sigma = 5.67 \times 10^{-8}\,\text{W/m}^2\text{K}^4\,,$$
$$\alpha = 0.96\,,$$
$$\phi = 800\,\text{W/m}^2\,,$$

we obtain $T = 341$ K for the maximum temperature of the water.

As

$$T \propto \alpha^{1/4}\,,$$

when $\alpha' = \alpha/2$, we have

$$T' = (1/2)^{1/4}T = 286\text{ K}\,.$$

Thus if the absorptance dropped by 1/2, the maximum temperature would be 286 K.

3026

Estimate the intensity of moonlight on earth.
(a) In watts per cm^2.
(b) In photons per cm^2 per second (in the visible range).

(*Columbia*)

Solution:

(a) Assume that the sunlight falling on the moon is totally reflected by the moon's surface and that the intensity of the sunlight on the moon is the same as that on earth, which is I, the solar constant,

$$I = 1.4 \text{ kW/m}^2.$$

Let r be the radius of the moon, R the distance between the moon and the earth. The flux of sunlight reflected from the moon per unit solid angle is $2\pi r^2 I / 2\pi$. A unit area of the earth's surface subtends a solid angle $1/R^2$ at the moon. Hence the intensity of moonlight on earth is

$$I_\text{m} = I(r/R)^2 \approx 3 \times 10^{-6} \text{ W/cm}^2,$$

taking $r = 1.74 \times 10^3$ km, $R = 3.8 \times 10^5$ km.

(b) The energy of a photon is $h\nu = hc/\lambda$. For the visible light, $\lambda \approx 5000 \overset{\circ}{\text{A}}$, and the number of photons arriving per cm^2 per second is

$$I_\text{m}/(hc/\lambda) = 7.5 \times 10^{12} \text{ cm}^{-2}\text{s}^{-1}.$$

3027

Estimate the efficiency (visible watt/input watt) of 2 of the following light sources: gas light, incandescent bulb, light emitting diode, fluorescent tube, mercury of sodium arc, laser.

(Columbia)

Solution:

The luminous efficiency of a light source is given by the ratio of the amount of energy radiated per second in the wavelength range $4000 - 7600 \overset{\circ}{\text{A}}$, which is commonly accepted as the visible range, to the total quantity of energy consumed by the source per second. Besides energy loss by conduction and convection, only a small fraction (the radiant efficiency) of the radiated energy is in the visible range. The luminous efficiencies of some common light sources are as follows:

gas light 0.001	mercury arc 0.08
incandescent bulb 0.02	sodium arc 0.08
fluorescent tube 0.09	laser 0.003–0.4 .

3028

Choose proper values from the given values below:
(a) The critical angle of total internal reflection in water is

$$5°, 20°, 50°, 80° .$$

(b) Energy flux of sunlight at radius of earth's orbit is

$$10^6, 10^2, 10^{-1}, 10^{-5} \text{ watt/cm}^2 .$$

(c) How many electrons per second flow through the filament of a light bulb?

$$10^{10}, 10^{15}, 10^{19}, 10^{25} .$$

(Columbia)

Solution:

(a) For water, $n = 4/3$, so the critical angle $i_c = \sin^{-1}(1/n) = 48.6°$. Hence the answer is 50°.

(b) 10^{-1} W/cm^2.

(c) The wattage of a bulb is $W = IV$. With $W \approx 100$ watts, $V \approx 100$ volts, $I \approx 1$ amperes. The number of electrons flowing per second is

$$\frac{I}{e} = \frac{1}{1.6 \times 10^{-19}} \approx 10^{19} \text{ s}^{-1} .$$

3029

Give a description, including the physical principles involved, of the following optical systems:

(a) A lens or mirror arrangement for converting a 10 cm diameter spherical 1000 candle-power light-source into a 10^6 beam candle-power search light. (Neglect lens or mirror light absorption.)

(b) An instrument capable of producing circularly-polarized light and of analysing light into its circularly-polarized components. (You may assume that the light is monochromatic).

(c) A system (in either the visible or radio-frequency range) for estimating the size of a distant (stellar) source of radiation.

(Columbia)

Solution:

(a) A lens or concave mirror would converge light from a source placed near the focal point into a parallel beam as shown in Fig. 3.22.

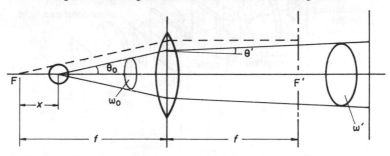

Fig. 3.22

Take the luminous intensity per unit solid angle of the source as $I(=$ 1000 candle-powers). Place the source near a lens or a concave mirror, which converges light within solid angle ω_0 into a beam of solid angle ω'. The beam candle-power of the search light is

$$I\frac{\omega_0}{\omega'} = I\left(\frac{\theta_0}{\theta'}\right)^2 = 10^3\left(\frac{\theta_0}{\theta'}\right)^2 = 10^6,$$

giving

$$\frac{\theta_0}{\theta'} = 31.6.$$

θ'/θ_0 is the angular magnification. By geometry

$$\theta' f = \theta_0 x,$$

and we have

$$x = \frac{f}{31.6}.$$

x is the required distance between the source and the focal point of the lens. Placing the source either to the left or the right of the focal point would produce a 10^6 beam candle-power search light.

(b)

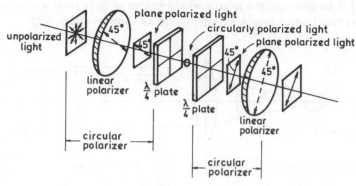

Fig. 3.23

Fig. 3.23 shows an arrangement consisting of a linear polarizer and a quarter-wave plate placed in the beam sequentially, with their optic axes oriented at an angle of 45°. When a beam of unpolarized light passes through the polarizer, it becomes linearly polarized. As its electric field vector makes an angle of 45° with the optic axis of the quareter-wave plate, the light emerging from the latter is circularly polarized. Beyond this, the arrangement consists of a $\frac{\lambda}{4}$-plate and a linear polarizer for detecting circularly polarized light. After passing through the $\lambda/4$-plate, the circularly polarized light becomes linearly polarized. Rotating the linear polarizer that comes after it, the emerging intensity changes from maximum to zero showing the linear polarization of the light.

(c)

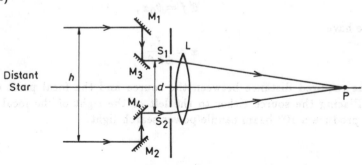

Fig. 3.24

In Michelson's stellar interferometer (Fig. 3.24), two plane mirrors M_1 and M_2 reflect light from a distant star onto two inclined mirrors M_3 and M_4 respectively. These in turn reflect the two parallel beams of light into a telescope objective after passing through the small slits S_1 and S_2 of separation d. Each beam forms an image of the star in the focal plane, and fringes are seen crossing the diffraction disc of the combined image. The visibility of the fringes depends on h. As h is increased, the first disappearence of fringes occurs at

$$h = 1.22\frac{\lambda}{\alpha},$$

giving the angular size α of the star disk.

3030

Metallic diffraction grating.

Purcell (Phys. Rev. **92**, 1069) has shown that electromagnetic radiation is emitted from a metallic diffraction grating when a beam of charged particles passes near and parallel to the surface of the grating.

The velocity of the charged particles is v and the grating spacing is d. Consider only the light emitted in the plane perpendicular to the grating and containing the particle trajectory.

(a) How does the frequency of the emitted radiation depend on direction?

(b) Suppose the metallic grating is replaced by a transmission grating, e.g., a piece of glass with grooves ruled onto it, still d apart. In what way, if any, will this change the emitted radiation?
Explain qualitatively.   (*Wisconsin*)

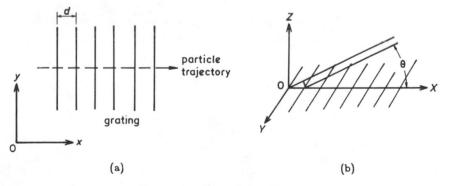

(a)   (b)

Fig. 3.25

Solution:

(a) The passing charged particles induce the electrons inside the metallic grating to move. The motion is limited by the very fine wires of the grating and the electrons are decelerated, which causes bremsstrahlung. Consider radiations emitted in XOZ plane (see Fig. 3.25) making an angle θ with the normal to the gratings. The radiations emitted from any two adjacent wires would have a delay in time of

$$\Delta t = \frac{d}{v} - \frac{d\cos\theta}{c} .$$

When $c\Delta t = m\lambda$, where m is an integer, the radiations interfere constructively. Thus the frequency observed at θ is given by

$$\lambda(\theta) = \frac{d}{m}\left(\frac{c}{v} - \cos\theta\right) .$$

For $m = 1$, $\lambda(\theta) = d\left(\frac{c}{v} - \cos\theta\right)$.

(b) As electrons in glass are not free to move, no bremsstrahlung takes place, i.e., no radiation is emitted.